HARVEST OF HOPE

HARVEST OF HOPE
Family Farming/ Farming Families

LORRAINE GARKOVICH
JANET L. BOKEMEIER
BARBARA FOOTE

THE UNIVERSITY PRESS OF KENTUCKY

This book is dedicated to the generations
who have been and will continue to be
a part of the living legacy that is Basin Spring Farm
and to all the other families who live and make a living
on Kentucky's farms.

Copyright © 1995 by The University Press of Kentucky
Scholarly publisher for the Commonwealth,
serving Bellarmine College, Berea College, Centre
College of Kentucky, Eastern Kentucky University,
The Filson Club, Georgetown College, Kentucky
Historical Society, Kentucky State University,
Morehead State University, Murray State University,
Northern Kentucky University, Transylvania University,
University of Kentucky, University of Louisville,
and Western Kentucky University.

Editorial and Sales Offices: Lexington, Kentucky 40508-4008

Library of Congress Cataloging-in-Publication Data
Garkovich, Lorraine.
     Harvest of hope : family farming, farming families / Lorraine
Garkovich, Janet L. Bokemeier, Barbara Foote.
          p.   cm.
     Includes bibliographical references and index.
     ISBN 0-8131-1921-9 (cloth : alk. paper)
     1. Rural families—Kentucky—Case studies. 2. Family farms—
Kentucky—Case studies. 3. Kentucky—Rural conditions—Case
studies. I. Bokemeier, Janet L. II. Foote, Barbara, 1945-   .
III. Title.
HQ536. 15.K4G37 1995
306.85'09769—dc20   95-16689

Contents

Preface

This book represents a harvest of hope. Nearly ten years ago we began talking about the need for a book that would present farming and farm life in the words of those who live it. Much of what most of us know about farming and farm life comes from brief visits or the memories of relatives who once lived on farms. Like old photographs, these images offer snapshots of farm life that are frozen in time. But the reality is quite different. Today's farm families face a world of change, and their lives reflect that dynamism.

Unlike other books on farming and farm families, this is a unique look at contemporary farm family life through the eyes of farmers and largely in their words. Most other books view farming as a business either in terms of the technological innovations that are changing its nature or in light of new policies directed at making farming more competitive and profitable. Most other works on the modern American farm assume a rational economic perspective that seems to deny the fundamental importance of family relationships and family heritage in shaping the nature of the farm business. This book rests on the belief that, to understand the nature of the farm business, it is necessary to understand the stories of the families that operate the farm businesses. Our challenge as authors has been to build an analysis of contemporary farm life that would capture the complexity, diversity, and dynamic nature of farm family businesses. It is an analysis that highlights the contradictions, mixed blessings, myths, and realities of a business and a way of life that have a powerful hold on the American imagination.

The three authors met when Barbara Foote interviewed Janet Bokemeier and Lorraine Garkovich about an earlier study of farm women. Then, as now, Jan and Lori were professors of rural sociology. They have spent their careers in agricultural experiment stations, conducting research and working as extension specialists. Jan has conducted research on families living in both metropolitan and rural areas. Lori has worked with rural communities and families as an extension specialist and has done research on family and community change. At the time, Barb was working as a freelance journalist for the *Louisville Courier-Journal* and writing a feature on farm women.

What was to have been an hour interview stretched into nearly three. After we commented on the multiple roles of women on the farm and in their off-farm jobs and the stresses these many demands created for the women and their families, there was a look of shared understanding on Barb's face. She began to talk about how she and her family had struggled with the need to find some kind of off-farm income to make a go of it and how this had affected them: "I thought I was the only one who felt this way, that we were the only family that had to go through all this." There was a sense of relief that she was not alone. We all began to believe that Kentucky farm families shared experiences and viewpoints that transcended the particular characteristics of their farm operations. And we believe that farm families throughout the nation would nod their heads and say, "Yes, this is how we live, work, and feel."

Although the farming experience is different in many ways from the work done by most Americans, we also believe that farm families' day-to-day lives mirror in fundamental ways the experiences of families in small towns, suburbs, and major cities. All families struggle with the pressures that come from a social world changing so rapidly that the guidelines and rules and expectations of yesterday do not apply today. All families share the pressures that come from struggling to do too much in too little time. All parents worry that they may not be doing enough for their children or that they are expecting too much of their children, and they wonder what kind of family and community heritage they will be passing on to them. All couples must contend with how to share the work and decisions that are necessary to keep a family together and strong in a changing world. And many families work in multigeneration businesses that require the extra effort to balance the needs and desires of an extended family with the demands of a modern business. The reader will come to realize that in many ways farm family life is a mirror reflecting the dreams, hopes, fears, struggles, and triumphs of many other American families.

Barb's story of life on Basin Spring Farm, owned and operated by the Foote family for several generations, is the thread that ties together the stories of the eighty other farm families who shared their stories with us during the summer and fall of 1988. The names of their home counties have been fictionalized to illustrate the essential character of their commodity base.

The poems are Barb's reflections on and responses to life on the farm and in rural America. Barb's accounts of life on Basin Spring Farm begin each chapter except chapter 1, introducing the stories that are shared by other farm families. Jan and Lori's analysis provides an interpretive frame for these stories, without, we hope, altering the farm families' intended meanings.

What you will read will often sound like a neighbor talking. The eighty Kentucky farm families represent a group portrait of American farm families. As in any group picture, we see similarities that show that the individuals share a common heritage, as well as the differences that make each individual unique. They share some values and beliefs, but often the meanings and consequences of even these shared ideas are different for the various individuals and families. The families are different in the ways that they face the risks inherent in farming as a business. Some embrace the challenge, the gamble, the pitting of themselves against nature and the odds. Others worry and look for ways to bring certainty and security to their business ventures. But all of them must make choices. Every day there are decisions to be made that will affect their businesses and their family lives. Excerpts from their stories as told to us are set off in italics in the chapters ahead. Farming, like all life, is about trade-offs, weighing the costs and benefits of different strategies and then accepting the consequences.

We want to thank our families and friends, who have had great patience and tolerance as we have worked through the development of this book. But most of all we wish to thank the Kentucky farm families who were willing to sit and talk "a spell" with strangers, usually for three or four hours. Their honesty, their openness, and their kindness has enabled us to step into their lives for a moment. We hope that we have given a fair account of what it means to be a farming family in Kentucky.

Prologue

I believe that imagination is stronger than knowledge—
   That myth is more potent than history.
I believe that dreams are more powerful than facts—
   That hope always triumphs over experience—
That laughter is the only cure for grief.
   And I believe that love is stronger than death.
                              —Storyteller's Creed

BASIN SPRING FARM

Farmers are storytellers, for their lives are the stuff of stories and myths. These stories and myths are what my dreams have been made of. Some have been dreams realized, some have been mirages, and others are still imagined as our lives change, as we go though our life at Basin Spring Farm. At the heart of my dreams have been these beliefs, real and imagined:

A family farm is where you raise chickens and have fresh eggs every day. *One of my overriding impressions when I first visited Basin Spring in the summer of 1967 was the hen house. It belonged to Mary, my future mother-in-law, and it stood just beyond the yard between the house and the barn and caught the morning sun, which suited its status as a center of activity after the rooster crowed each morning. After breakfast the chickens would have to be let out to do their chicken work of hunting and pecking for bugs and insects and other wonders that went into making rich yellow yolks and firm, clear egg whites for the angel food cakes (the whites) and lemon chess pies (the yolks) that Mary would fix on Friday afternoons when we rolled in from Lexington.*

But most of all I enjoyed poultry life, the background clucking of chicken activities around the hen house, the routine of putting them up at night and letting them out in the morning and seeing that they had water and feed. And many a time, in some small hour of the night or early morning, I would hear Mary call out to her son, "Jimmy, something's in the chickens!" and Jim would grab his .22 and bolt for the door, where, sure enough, a varmint of some kind—fox or raccoon or weasel—had been bothering Mary's brood.

The fall we moved to Basin Spring, Jim and his father built a new henhouse in one end of the old corn crib, since the old hen house had been torn down when the new shop was built. So the chickens were housed before we housed ourselves. Several years later, after we had bought the farm from Jim's parents, Mary and Gerard, Jim surprised me with Araucana chickens that laid pastel eggs. I was enchanted! They were pale green and blue and yellow and pink, and I gave the children an old Easter basket, which seemed appropriate for gathering a new color combination every day. But in the late winter of 1981 a varmint wiped them out, probably a weasel that burrowed nightly under the hen house walls. As fast as I plugged up one hole it would dig a new one, leaving me with fresh casualties the next day.

This went on for several weeks, I cried when it got the rooster, and I saved a few of his tail feathers, which are still in a bottle on my desk. It was the end of the Araucanas, and of the chicken era. I was beginning a job at the newspaper thirty miles away, which made it impossible to be vigilant. We needed something more secure than the old corn crib, and we needed a yard to keep the chickens confined, neither of which was cost-effective at the time.

Today we seldom eat eggs, maybe once every week or two on Sunday mornings when everyone is home and we fix a big breakfast. One day, how-ever, I will have Araucanas again, for the pure pleasure of them. And per-haps there will be grandchildren and great-nieces and -nephews to gather eggs. But there will be chickens about, clucking and scratching and doing their chicken work. You can have a farm without chickens, but I would rather not.

A family farm is where you grow a garden. *I started with flowers and herbs and a few greens, then graduated to tomatoes, green beans, squash, pep-pers, pumpkins, cabbage, and broccoli. Some of my favorite things to grow are variegated sunflowers for the birds and Indian corn and pumpkins in honor of fall.*

My sister-in-law, Linda, introduced me to canning when I helped her put up green beans the fall I was waiting for Jim to return from Vietnam. Seeing the sparkling jars filled and sealed gave me one of the most tangible feel-ings of accomplishment I had ever enjoyed, and I was hooked forever on the merits of canning. When we moved back to Basin Spring, I canned every-thing—corn, carrots, beets, green beans, tomatoes, pickles, sauerkraut, and squash. The squash was not too successful, but I canned it anyway and proudly lined my jars on shelves in the cellar.

Since I have been working off the farm, however, time is at a premium. I know women who stay up until midnight or one o'clock in the morning wait-ing for the canner to cool, then go in to work the next morning at seven thirty, exhausted and bleary-eyed. Not me. In recent years I have done more freezing than canning—even tomatoes, which I wash and core and freeze whole in plastic bags. Although freezing is more expedient, the accomplishment does not quite seem the same as admiring row after row of colorful, sparkling jars filled with your own vegetables for the winter. It is just another of the ways I feel that I am here but not here.

A family farm is where you milk a cow, churn butter, and raise baby calves and all your own beef. *Her name was Bluebelle—or Baybelle, as our son, Gerard, called her when he was learning to talk and would toddle to the barn with his grandfather, who milked her by hand twice a day. Mary would strain the fresh milk into gallon jars and refrigerate it immediately, so we enjoyed the abundance of fresh raw milk and consumed it lavishly. When the cream had risen to the top, I would skim it with a gravy ladle into jars, which we kept refrigerated until there was enough of the heavy, sweet cream to fill a one-gallon Dazey churn to two-thirds full.*

I loved to churn. It was easy, but you had to be mindful of a few things. The cream had to be sweet, for any off-taste would cause the butter to turn sour. You could not fill the churn too full; and the cream had to be just the right temperature, which we gauged with a dairy thermometer. Allowing for these factors, you could churn butter and have it ready to work in ten minutes. Sometimes I saved the buttermilk, but often I gave it to the cats and dogs. Then I would rinse the bright yellow butter over and over, pressing out drops of buttermilk with a wooden paddle until the water ran clean. I would salt the butter lightly, roll it into balls, and wrap the balls in cellophane for the freezer.

We stopped milking at Basin Spring several years ago. Mary and Gerard grew tired of it, and Jim and I didn't have time. When the children started school they acquired a taste for margarine and pasteurized milk. So that is how it went. The gallon jars are now used for sun tea or large collections—buckeyes, rocks, or whatever a teenage boy brings in from the woods and fields. The Dazey churn is filled with pinecones and decorates a shelf. I treasure these things. And I wonder, will I use them again?

A family farm is the best place to raise children. *Our children have virtually grown up at Basin Spring. They have enjoyed space and privacy and a countryside to explore, to learn the laws of nature. They have known grandparents and cousins and aunts and uncles. But most of all Basin Spring has given them room to dream.*

At times the price seems high, however. Our children have watched their parents squeeze two full-time jobs into what was more than a full-time job and a way of life. This has not always been an example of how to live well. In fact, we have become people very typical of our times, meeting ourselves coming and going with too-much-to-do-and-too-little-time, "country style." It is the antithesis of the life that drew me here twenty-four summers ago, and it has made me wonder whether our gains have been worth our losses in energy and time—time to work well, time to enjoy our work, time to relax, and time to play.

*The compromises can be great in today's agriculture, the alternatives for survival neither ideal nor idyllic. But I would say, **yes**, the farm is the place to raise children. The youth who has learned to pitch hay or haul tobacco in ninety-five-degree heat and 90 percent humidity (standard conditions for an Ohio Valley summer), who has worked the long rows in a tobacco field, hoeing and topping and cutting tobacco, and then has stripped it with fingers stiffened by November winds and **finished the job** has learned a lifetime about work and patience and time. That is an invaluable lesson for any age. But it is not a life for everyone. First you must love it. Then you must have energy and stamina and resolve—more than enough. For living the farm dream will take all of you. And more.*

A family farm makes a strong family and a strong community. *If we were all like the Waltons I would say unequivocally, **yes**. But we are not, so my answer is **yes** and **no**.*

During our years in Pennsylvania, the Mennonites and Amish set a standard for family and community that is unrivaled in our culture. To watch a gathering of Mennonite men raise four walls and roof a barn on a spring day while the women serve lunch and quilt under the trees is the quintessential experience of a farming community, a culture self-contained, in which people take care of one another.

At Basin Spring, we have found ourselves isolated within the constraints of time and energy. Escape from the demands of work and farm has often meant curtailing time with family and friends. Though we do swap work and equipment with our immediate neighbors, more than once I have thought to myself or overheard someone say, "It is a shame we have to run into our neighbors at the funeral home." It is sad, but too often true. This is not "neighboring," nor is it culture. It is unsound living that does not build, but fragments; it does not sustain, but at best maintains. It does not leave much time or energy for interaction.

*There is a way to survive **and** to live in rural culture today. Though we have not figured out the terms that can wholly sustain us at Basin Spring, I feel we will have to return to some of the "old" ways. A way of living that is* *not old, but sound. But we will have to eliminate some things, and we will have to slow down.*

A family farm is where the seasons unfold. *A friend of mine says it best. After taking a full-time job off the farm, she laments the loss of time to cook and clean and keep house the way she is accustomed to doing, let alone time to enjoy where she lives. "The seasons on the farm are slow," she says. "Now it's just a hurry-up pace. The summer's gone, the winter's gone. You don't relish the seasons the way they used to be."*

The way it used to be at Basin Spring: a snowy day meant a rest after the cattle were fed and the fires were fixed. Now a snowy day is an ordeal to drive through twenty-five miles to work. And a rainy afternoon is just another afternoon, not a time to let down from pushing in the hayfield or the tobacco patch.

There is no built-in time off any more, because when you work off the farm, you still work on the farm. Which is to say that **you work all the time;** *you are always* **on the job.** *That is not necessarily a good way to live, but it is the way many farmers operate today, farmers like Jim for whom no price is too high to keep a family farm alive.*

On a family farm, hard work will be rewarded. *Not necessarily. There was a time, a far simpler time, when a farmer was practically guaranteed a living by showing up in the marketplace with something to sell. But rural America no longer operates a simple economy. Ours is a multidimensional global economy that responds to the Russians invading Afghanistan and a rise and fall in the price of gold. Oddly, the people who seem the farthest removed from world events in their day-to-day work are among the few most directly affected by it—the farmers and farm families of America, who plow and plant and harvest regardless of world politics. So the game plan today is not to work hard but to work smart. You must know production, know the government programs, and know the market. It takes as much wit and wisdom as it ever did to make it today. Plus a vigilant eye on the world.*

A family farm is the bedrock of America. *This is no myth. Agriculture undergirds our entire economy. And there is no more graphic way to experience this fact than to spend a day driving across the country through the corn, wheat, and ranch country of this land, or to behold from the air the patchwork landscape of fields and farms that comprise our rural economy.*

Farmers, it seems, are the first to feel the reverberations of economic downturns and are the last to recover. The Great Depression was felt across

*rural America long before Black Tuesday left Wall Street in tatters in 1929.
The last recession hit Basin Spring during the winter of 1978-79, several years
before the word* **recession** *was breathed by economists. And we have yet
to recover when corn is bringing less per bushel than it did in 1973.*

*But the American farm is more than an economic bedrock. It does, in
fact, embody family, community, and conservation. These things are not
trendy; rather, they are lasting enterprises—commitments, loves, that which
has always been healthy about American culture in general and rural culture
in particular. But our farm economy is not healthy and cannot be healthy
and viable until the market will support the cost of production for the food
and fiber producers of this country who, in turn, support farm laborers, the
agricultural businesses, and the communities that make up rural America.*

We have our work cut out for us.

TWICE RINSED, TWICE BLESSED

After hanging out two loads of wash
    between showers I sit down at my desk.
"Tell the story only you can tell," I am advised
    at the women writers conference
and I wonder—what is the story of farm women?
    Who are we today? What are we? I do not know
all I can say is
I love
the sound of rain on leaves
seeing the light change
    from morning to afternoon
    in these old rooms. And I love
the whir of the ceiling fan
birds twittering in the trees
zinnias blooming in the garden
    butterflies in the zinnias
squash, peppers, Italian plum tomatoes
spinach in the spring, mustard in the fall
    and pumpkins and straw flowers and ornamental corn,
cattle lowing in the pasture, frogs singing at the spring,
dogs wagging a greeting
cats lying in the sun.
Given these things
wet laundry is
    just a second rinse

It is pouring now
      on the clothes, on the pasture,
      on soybeans and corn. Jim said
a good rain today could make us
      several thousand dollars.

I love
the wind blowing the curtains
the smell of rain
      turned earth, new growth
here
today.

This is my story.

1 Family Farming in Changing Times

Family farming is rural America. The image of rural America as the storehouse of the traditional values that built the nation—self-reliance, resourcefulness, civic pride, family strength, concern for neighbors and community, honesty, and friendliness—persists, as many recent surveys show. Farm families accept these images of farming and farm life as their own. Many farm families live on farms that have been farmed by their ancestors over many past generations. They base their images of farming and farm life on this rich tradition, and they have felt secure in these images. What we have found in our studies of today's farm families, however, is that changes in farming, in the agricultural industry, and in rural communities have begun to erode the security of these nostalgic images.

Flipping through the pages of history, we find different images of farm family life. One of these shows a family working together and relying upon common sense, hard work, generations of know-how, and *each other* to build a home, a business, and a future for their children. Neighbors and relatives come together in support of the farm family. Their community is close-knit and built on traditional values of independence, honesty, self-reliance, and the importance of family. Their economic and subsistence needs are more than met because their hard work on the farm produces the food bought in the cities. This is the image of the farm family and agriculture captured in the words engraved above the portals of the Union Station in Washington, D.C., a century ago. "The Farm. Best Home of the Family. Main Source of National Wealth. Foundation of Civilized Society. The Natural Providence."

Another image is of a family that is bent from years of harsh, unending labor, aged before their time. With limited educations and limited expectations, they scratch a living from the land. Their contributions to the national agricultural production are inconsequential, for they see their farm more as a way of life than as a business. They raise children who leave farming in search of better lives in the city. Vacant stores dot the main streets of their towns, towns left behind by economic progress and abandoned by their youths. Rural people's limited access to health care and varied jobs perpetuate high poverty and social distress. This is the image of the farm family and agriculture satirized by writers and bewailed by politicians.

Which of these is true? Is one the picture of an agricultural life that we once had and then lost, and the other the picture of what farm life has become? In fact, there is some truth in both images. At different points in time and in different places, each type of family farm and agricultural community has existed and still does.

Within the social sciences there is growing recognition that we are what one person calls "homo narrans," the species that tells stories. Our stories involve the selection of particular events from the past to which we attach meaning. In the telling of stories we are explaining why and how events happened, in terms that are meaningful to us and our listeners. Our stories link our private lives with the public world around us, giving a continuity to personal and social history and giving shared meanings to everyday life. Linguists argue that groups of people share a way of talking and storytelling that affirms they belong to a community of others who have a shared understanding of how people should behave, what they should strive to be, and their common responsibilities. Stories tell of our shared heritage, our collective understandings of our community's past, the events that shape our sentiments and interests in the present. Storytelling, as Barb notes in her prologue, is what binds the community of farmers together and gives meaning and direction to their lives. Yet when the shared understandings of the meanings of our stories begin to break down, conflict and dissatisfaction can emerge.

Barb's story of life on Basin Spring Farm is similar in many ways to the stories of the eighty other farm couples we interviewed, suggesting several common themes that bind the community of farmers. Yet farm couples of the older generation are less likely to admit that their lives, their realities, veer from the romanticized myths of what farming and farm life *should* be like. Younger farm couples are willing to tell stories that are more balanced, that admit that farming and farm life do not match the romanticized myths told by their parents and grandparents. It is this more analytical storytelling that comes clear in Barb's prologue. However, it is important to remember that because farm families share a community based on a way of life and a particular kind of business, the stories and myths about farming and farm life have a powerful influence on how they come to understand their world and how they adapt to change.

For example, farmers can wax eloquent on how their crops are growing, or the weather, or the cow that had trouble calving last night, but they become surprisingly reticent about their personal lives. That reticence became apparent during the farm crisis of the early 1980s, when farm families that everyone thought were doing well simply packed up one day and left behind the farm and home they had shared for decades. Or when the word spread that Joe D. had committed suicide because they were going to foreclose next week, and nobody knew he was in trouble. The independence and self-reliance that can be seen as an enduring theme of any story of life on the farm can also be a barrier, locking up feelings and concerns.

In the summer and fall of 1988 Jan and Lori interviewed eighty farm families in three Kentucky counties. Each county is identified by a name that reflects the most common commodities raised there. With the help of county agricultural extension agents, we identified farm families of different ages and in different types of enterprises at each site. A phone contact prior to arrival in their community informed the couple of the purpose of our research and our interest in spending an hour or two talking with them. Each family generously invited us into their home, where we spent anywhere from two to four hours listening to the stories of their lives on the farm. Sometimes we were met with enthusiasm and sometimes with polite reserve, but we always left with the request, "Please come back, it's been so nice to talk to someone who is interested in how we think and feel." And then came the question we had heard so many times before: "Are we different? Have others told you the same things?" There was a sense of relief when we acknowledged that we had heard their words before.

We cherish the memories of that summer and fall and the encounters that we will never forget: The mother and daughter who proudly gave Jan a tour of their flock of sheep, built with loving care over several years of a family partnership. The older couple who spoke in monosyllables throughout the first half hour of the interview but by the end, two and a half hours later, were standing beside the open car urging Lori to come back for breakfast the next morning. The sixty-ish farmer who, when Lori thanked him for his time, said, "Wait, I want to get something." He rummaged on the dining room table until he found the brown wrapping a magazine had come in that day. Putting on his glasses, he said, "I wrote a few things down that I wanted to be sure to tell you, and I want to check this list." He carefully read through his list and then said, "Oh, I forgot about this." And we will not forget the many kitchen tables we sat around, sometimes until midnight because every time we noted how late it was, they insisted we have another cup of coffee. The stories we heard are presented throughout the remainder of this book.

One of the most interesting discussions we had with the farm couples we interviewed revolved around their answers to the question, "What do you think people who don't farm believe about farming?" This always led to a listing of what they called "popular misperceptions" of farming, such as the belief that farmers only work a few months a year and then relax, or that farmers are uneducated. Yet present-day myths of farming generally are built around some kernel of truth. We did interview a few farm families, typically older couples operating grain farms with little or no debt, who would spend several months raising their crops and then, once the crops were sold, would take a few months off before the next growing season. And we did interview farmers, some very successful and others just holding on, who had not finished high school. The great diversity in farm families means that all of our myths about this community have some roots in reality.

This book weaves the story of contemporary farm family life by capturing the contradictions, mixed blessings, myths, and realities that underlie this business and way of life that have such a powerful hold on the American imagination and on the families who live this life every day. What have been the different types of farm households? How and why has the farm as an agricultural enterprise changed? What are the characteristics of new farm households? These questions will be examined in part through farm men's and women's own stories of their family businesses, their way of life, and the forces they see shaping their lives.

Why read about farm families? Farm families make up a very small part of the American population. Only 2 percent of the nation's population lives on farms. Yet a lot of people, in fact, about one out of every four Americans, live in rural areas, that is, in open country or small towns of less than twenty-five hundred people. Many of these rural areas are greatly influenced by the experience of local farm families, who make up about 10 percent of the rural population. It is the local farmers who buy trucks from local auto dealers, fertilizer from the feed store, and lumber and insurance from other local small businesses.

There are good and bad consequences of "progress" in agriculture. On the good side, mechanization has made the physical work of farming easier: much of the backbreaking stoop labor of production is now only a memory of how farming once was. Mechanization has also been the basis for the substantial increase in productivity on American farms. Rural electrification brought a new way of dairying, extended the workday for those who wanted it, and brought the comforts of an urban lifestyle to farm households. Government programs have provided some stability to farm income and have expanded markets, and government-sponsored research has offered new crops that have bred into them resistance to disease and pests as well as greater productivity.

On the downside has been a loss of farm jobs, jobs replaced through more reliance on machinery and technology. Throughout this century mechanization has reduced the demand for labor on farms, and many of the unskilled workers who once worked on farms have left in search of other jobs. As farm labor and farm families have left farming, many rural towns and villages have been hurt. A drive down the back roads of the nation finds boarded up Main Streets and abandoned farm homesteads, mute testimony to the interdependence of farming and rural community businesses. Furthermore, as the number of farms has declined and the population dependent on farming has diminished, many Americans have tended to see farming and farm families as increasingly marginal to our economy and rural community life. One of our respondents noted that the early '80s made many people revise this estimate. "I think this recession in farming has caused people to realize how much the small businesses around here depend on farming. We have seen banks fail and stores close because farmers couldn't pay their bills. So there's a new awareness of how much we depend upon agricultural income to make our economy go."

There is another way to think about farm families, one clearly illustrated by these stories. Farm families tell us much about families working and managing their own business operations. Small family-owned and family-operated businesses are thought of as the backbone of our economy. It is the American dream to become your own boss and to invest in the chance to make it big. Every year more than six hundred thousand new businesses are incorporated, the majority of them small, with fewer than five employees. Many of these businesses rely on the labor and commitment of family members to succeed. It is not unusual for parents to plan to pass the family business on to their children and to explore this possibility by having children work with them in the business, beginning at an early age. The hopes, challenges, and problems that confront family farm businesses are like those of hundreds of thousands of other small businesses across the nation. So a look at family farming helps us understand the changes that are transforming many rural communities, whereas a consideration of farming families provides insight into a major sector of our economy.

Kentucky is an ideal location for a study of contemporary farm life. Nearly all the types of farming that occur in the rest of the nation are also found there. From the western Delta to the south central counties, large-scale cash grain operations (500 acres or more) are typical, many enhancing production through irrigation. Middle-size general farms (150-500 acres) with a mix of livestock, corn, hay, and tobacco predominate in the central counties. To the north and east smaller general farms (less than 150 acres), often operating at subsistence level, are common.

Kentucky farm families are as diverse as all families in America. In Kentucky as in the rest of the nation, there are families who have operated a farm on the same land for more than 150 years and couples who have decided to break from their own families' past and begin farming as a career. Some families have no one with a job off the farm; in some, one spouse does work off the farm; and in most, both spouses have jobs away from the farm. Farm couples are old and young; they have grown children who have left the farm and grown children who continue to operate the farm with them, and they have young children. Single men and single women operate farms; there are brothers, fathers and sons, mothers and daughters, and grandparents and grandchildren in partnership on farms. Like other family-owned businesses, some farms are independently owned and operated by the family; others are operated by the family but producing under contract with major food processors; and some are a division of a national or multinational corporation. There are farms surrounded by expanding cities and others more than twenty miles from the nearest town.

Our frozen images of farming and farm family life mask the incredible diversity that marks contemporary farm life. Besides the diversity in types of farms just noted, families have developed diverse strategies for survival, which are grounded in and contribute to a variety of lifestyles. There is no typical farm family: the composition, structure, and family relationships are as heterogeneous as the types of farms they operate. Moreover, the values and expectations that give meaning to their lives and underlie their decision making are as different as those found on any residential street in America.

Within this diversity of farming, is there a commonality of life experiences? Over the months of interviewing we realized that despite the size or nature of the farm operation, despite the differences in family structures or lifestyles, we were hearing recurring stories, seeing recurring images, recognizing recurring themes. Farm families tell stories of the constant struggle with the elements and their growing concerns for the safety of themselves and their children as well as the safety of the commodities they produce. The images of families, often two generations working together, and of rural communities left behind; and the theme of being the last of a disappearing breed, struggling to maintain an endangered heritage, are threads that bind their stories.

Persistence and the commitment to farming for better or worse are common themes in their stories. There is a recognition that to succeed you need more and more formal education but that education is not enough to guarantee success. There is a sense that nonfarmers, and especially policymakers, do not understand what it means to farm today. Government policies and programs affect so many aspects of farming, but the bureaucrats who make these policies and administer these programs usually do not grasp the consequences of their actions for farming and farmers. Farming is a family business, they agree, but there are many tensions between the generations and within families. Strong feelings about what is proper for men and women to do, believe, and strive for shape farming as a business and a way of life.

Another recurring theme is the struggle over how to view change. Is change progress, or does it lead to a loss of those things we define as precious? Changes in technology have made work easier, but new machines have also ended the reliance on neighbors for help. Is the loss of communal work patterns at the gain of easier work a case of the glass half empty or half full? Increasingly, women work off the farm, bringing greater income to farm families but at the cost of the time farm couples once spent together. What do all the changes mean for the way we understand farming and farm life? How should we feel about these changes? How should we deal with them? This is the challenge of contemporary farm life.

Change is ever present on American family farms. The farm of today would be a wonder to the farm family of the early twentieth century, or even of thirty years ago. New technologies, new and more complicated machines, chemicals, hybrid seeds, bioengineered seeds, and superproductive livestock are the defining characteristics of today's farms. Social issues such as who is farming, how farm communities work, and the matter of balancing work on the farm, at home, and in other jobs are the problems of contemporary farm families, who struggle for solutions that their grandparents once took for granted. The '80s saw a rolling wave of farm crises sweep agricultural America. Using a Darwinian logic, many assume that farm crises are a natural market outcome: the weeding out of the weak and the noncompetitive. But the current crises have hit today's "young farmers of the year" hard. Farm families are like many others in the '90s who are finding that having a college degree or being with a firm for twenty years does not spell security or insure success. Knowing good farmers who do not succeed leads to incredible uncertainty and stress related to whether *you* will make it. Thinking back over the last ten to fifteen years led one farmer to say: "A lot of people try farming and fail, and they think it's their fault because they fail. It's not always that way. You can do a fine job in farming and still lose it." This is perhaps the best summary of how things have changed in farming. Hard work is no longer enough to guarantee survival in a changing world.

People have certain expectations about their standards of living and ways of life that are based on past experiences and on comparisons with others in similar situations, as well as on what we have come to consider our rightful due as free and hard-working citizens. But when families find themselves with life situations that are no longer like those of the families they grew up in, or when they are at work where they do not feel in control or able to predict their chances of getting ahead, then they cannot count on traditional ways to help them know what to do or expect. Instead, they must rely more on creative strategies and a general optimism toward the future.

The changes that have occurred in farming and in America bring new challenges. Customary ways of doing and thinking about what we do may no longer apply. What used to work and lead to success does not always fit the changing world. Grandparents sold their goods on a regional or national market where knowledge of droughts in Russia or a good harvest in Canada was not necessary to survive. Today, success depends on knowing how the North American Free Trade Agreement (NAFTA) will affect import controls and how the General Agreement on Trade and Tariffs (GATT) will affect the price of American grain in Europe. This climate of constant change creates considerable stress. Too much seems to be happening, too many events are interacting to be able to get a good grasp on how these changes may affect you or what you can do in response.

Responding to these changes requires a high-wire balancing act. The shift to a world market and new technologies, along with economic and social changes in nearby communities and the quickening pace of life, introduces enormous stress. You feel like Alice in Wonderland, trapped in a world where the rules are constantly changing and you must run faster just to stay in place. One thing that helps you survive is hope.

Hope, optimism for the future, a firm belief that things will be better next year, are themes that weave through most farm families' descriptions of their lives and work. In a sense, each spring begins with a harvest of hope. You must be an optimistic person to spend all the money necessary to get a crop out in the spring with a firm belief that several months later you will harvest a crop that sells for enough to recover your costs and put some aside. To buy feeder calves in the spring or to breed your sows you must have great confidence that no illness will occur and that the price of feed will not skyrocket or the market plummet, so that months later you can sell and cover your costs with some left over. And it takes overarching faith to plant or breed this spring when last year, and maybe even the year before, nature and perhaps the market conspired to produce net losses. Yet this kind of hope is precisely what keeps farm families going year after year.

But such hope and faith are not without cost. Many farm families often feel "stressed out," full of uncertainty and anxiety, and it is not surprising. Faith alone does not buy food for the family or pay the utility bills, and no lenders will accept hope in payment of loans. Although the firm belief that things will be better next year seems to sustain the commitment to farming, there is also a recognition that faith alone is not enough. Farm families are constantly in search of strategies that will give them more security in times of economic uncertainty and social upheaval. Many look to older strategies, ones used by parents or grandparents or earlier generations. Yet in times so remarkably new, how can tradition give support? It is the constant pressure to adapt to changes, most of which occur for reasons beyond our control, that leads to the stress and uncertainty in farm families and challenges their commitment to traditional ways. They rely on hope to get them through.

For many farm families the defining characteristic of their lives is adapting to constant change in an endless search for the right combination of strategies that will bring them success. Many feel that rational planning, logic, and exceptional decision making, although necessary, may not be enough in a world shaped by unpredictable weather and the influence of a host of other conditions outside the control of the individual farm manager. Instead, there is the hope that they can find a way of living that gives security in times of economic uncertainty and social upheaval. This hope is grounded in the strength of the family, which shapes farming as a business, and it is hope that shapes the family's life on the farm.

One young woman who had been raised on a farm and was very certain she would not go back to the farm after college recounted an incident that often happened on the farm: "When I was living at home people would call Grandpa and say, your cows are out. This could be at any time of the night and of course we'd all get dressed and go out and round up the cows. Do you know how hard it is to find a black cow at night? I mean, we walked one night for three hours trying to find three cows. We

finally found two of them in the middle of the road, but we couldn't find
the other. And I felt, 'Give it up! After two or three hours the cow will
come home! And if it don't, I don't care, someone else can have it!'"

In many ways today's family farmers are like this family searching for cows
in the dark of the night. Often they are called upon to make decisions or
react to problems in the business without full information and without
the means to control many of the factors that will influence the results of
their decisions. They feel isolated and alone in their struggle to make a go
of farming. Tensions in the business spill over into family life, because
the business is where they live and their labor is supplied by family members. And sometimes, even after putting in hours of hard work, they find
that the effort has been wasted. Tried and true solutions that served their
parents well no longer work or no longer work as they once did. There are
many questions, but the answers often are hidden in the dark.

The family is an essential component of farming. Family farming is
about family for several reasons. Most workers on the operation are related by blood or kinship, and the farm, that is, the land and home, is
part of one's family heritage. The family farm is more than soil and livestock. It is also traditional strategies for how to farm, care for, and use the
land and traditional meanings and values attached to the land. The family farm practices are passed from one generation to another with the
gradual transfer of farm decision making from the parent to the child
and through children bringing their families to live on the farms where
they were raised.

Unfortunately, time-honored family farming practices and attitudes
are becoming outmoded or simply incapable of responding to the rapid
changes. Most people who farm today believe that a college degree is essential for a chance at success. And many decisions that farm families make
are influenced by their potential effects on the family's ability to send a
child to college. This is only one example of how farming in a family context affects farming as a business. Ironically, many youths who go to college intent on returning to the family farm find other opportunities with
much better chances for economic security. They never return to their
family farm. Or if they do return to farming, they may find that they are
not able to earn what their college classmates can earn at nonfarm jobs.
They often find that all the formal or technical training helps little when
the rains fail to come or the bottom falls out of the grain market.

Farming is a special type of family activity. Because of one's membership in the family, certain expectations are made about what one will or
will not do at home. In general, farm children have certain chores and

farm tasks that are their regular responsibilities. Unlike the typical child-hood chores, these tasks have significant economic consequences for the family business. Children know that they are contributing to their family's business, and indeed, the success of that business is partly attributable to how well and how responsibly they do their tasks. The people one works with on the farm, one also plays with, eats with, and shares intimate moments with.

Thus, farm work permeates all aspects of family life. It has always been this way, but today's world is different. Organized after-school activities beckon farm children, and a host of satisfying careers off the farm call to young adults. Some farmers look around at other workers and businessmen and feel that they have been left behind in the rising standard of living. Hard work, faith, and careful management helped their parents and grandparents reach a comfortable life, but the same ingredients seem not to produce the same result today.

Farmers have always had to cope with great uncertainty as to weather conditions, market responses, animal and crop diseases, pests, and weeds. For modern farmers we can add to the list government programs with a history of changing eligibility requirements and political purposes, world market crises, waning rural communities, and an explosion of new technologies. In this atmosphere of uncertainty farmers must make decisions with long-range consequences, even though they have only a short-range sense of risks.

For example, a dairy operation faces the decision of either building a new milk parlor and barn or selling out their herd. Milk prices have fluctuated wildly since the milk price support system was begun in 1987. Technological advancements, such as the recently approved bovine growth hormone, will have impacts on the dairy industry that no one can predict. In parts of the country the rise of megadairies, milking fifteen hundred to two thousand head a day, are creating a new price and management environment for more traditional family dairies. And then there are the hints of tightening regulations spurred on by environmental concerns and an increasingly vocal animal rights movement. In the '80s many dairy farm operations did very well financially. But will they continue to do so? It is the uncertainty surrounding the answer to this question that makes farming today a search for cows in the dark.

Farming is an uncertain business. One farm husband summed up his feelings about farming this way. "Oh, I like farming. It's just that I can't make a living at it."

In terms of health and safety farming rates in the top three high risk occupations. In economic terms farms are at the mercy of natural phenomena—weather, soil conditions, diseases of plants and livestock, and natural disasters. Farming is also influenced by a marketplace that responds to land prices, the value of the dollar, politics, federal policies, and global competition. Farmers find they are constantly affected by forces over which they have little or no control. To some extent farmers learn to compensate for the natural limits. They know that certain crops cannot possibly flourish on their farms with their particular climate and soils. They, like people in nonfarm occupations, invest in insurance or have jobs on the side to give them a sense of security.

A lot of people have to deal with problems that are not of their making, or find that they have no control over the factors that influence their businesses. When General Motors closed its plants in Michigan in the early '90s, many waitresses and clerks in local restaurants were laid off and their bosses went out of business. Why were they out a livelihood? Well, it was not because of their lack of hard work or dedication. Such elements of uncertainty or risk are always present.

Farming is a risky way of life, but it has been that way for centuries. At times when things go wrong, farmers lose faith in past strategies for farming and change their approaches. Changes in one's business and one's family can bring both a sense of loss and a sense of accomplishment. This is just one example of the ambiguity and contradictions experienced by small-business operators. We found that farm people do not all look at farming the same way. Some see their businesses and ways of life fraught with risks, but others see the uncertainty as a challenge; some like the gamble, and others want more control. "To me risks and challenges go together. I consider my risk as a challenge. That is what I love about it. I love the challenge of not knowing what my paycheck is and realizing that the better job I do the bigger my paycheck is going to be. I love it like that. It keeps me alert because I know that when I goof up or when I slack back, my paycheck will be less. I think it is a challenge to a young person. I think it causes you to do your best. It certainly causes you to depend on yourself. You have to know a lot of things to be a farmer. I don't say that with any thumbs under my lapel. But a farmer has to be a scientist and he has to know medicine. He has to know a little about astrology. You've got to know plants. You've got to know your chemicals. When we first started using chemicals, we just needed to know what would kill a weed and what would kill a fly. But now they have gotten so hazardous, and we've got to know which one we can use on the tobacco if we're going to work in it. I count it a challenge. It is a continuous educational thing. You learn every day and every year."

One of the themes running through farm families' accounts is the difference it makes whether a family thinks about its operation as a family way of life or as a commercial business enterprise. Advocates of both perspectives see farming as a means to earn a living. However, they differ in their emphasis on profit maximization in decisions, on inheritance and legacy issues, on reliance on family labor, and on embracing change and risks.

How do families deal with the risks and uncertainty of farming? In talking with farmers on dairy operations, large cash-crop operations, beef and pork farms, and small diversified farms, located on large spreads with thousands of acres and on small acreages that some agricultural scientists do not even consider farms, we found a mix of strategies and answers. Many farmers build strategies around a sense of mastery and trust in the land and in God. Most have financial plans that, if all goes well, will let them turn a profit. When things do not go well, as with the drought of '88, the farm family may accept the setback and plan for their next crop, or leave the farm for work elsewhere, or look to friends and relatives or to their own sense of humor to help ease the disappointment. As one farm husband told us: "A farmer is the only guy I know of that keeps on taking a beating and always looks forward to next year. They are the only people I ever knew that could take a lickin' and keep on tickin'."

Making sense of the risks depends upon how you look at what you do. If farming is primarily a family way of life, then "getting by," a phrase we often heard, is enough, for it means you are providing for your family in the context of a desired lifestyle. For these families a crisis develops when they begin to believe that the rewards of farming as a lifestyle no longer compensate for lower returns on personal and financial investments in the business. It is a crisis of values and expectations that is not easily resolved. If farming is primarily a commercial business enterprise, then "getting by" is not enough. Getting by means that the conditions of business are so risky, so uncertain, so competitive that even the best managers have difficulty maximizing their returns. For these farm families, when the balance sheet shows losses year after year despite the most aggressive management strategies, the solution should be relatively simple: sell before your losses are so great that you lose all your assets. But even here a crisis of values and expectations may result, for despite the belief that the farm is *only* a business, the years of hard work together as a family make it more than that.

Another theme that emerged when talking with farm families is a sense that they are on a treadmill that is continually speeding up. The family wants to take advantage of better prices, so they expand the size of their farm operation. To cover the costs and the ever-present need for cash, one of them will take an off-farm job. Pretty soon they begin to feel tremendous time pressures: not enough time to get work caught up on the farm and not enough time to be with the family. As one farmer put it, "The hurrier I go, the more behinder I get." The sense of constantly juggling too many demands with too little time is a thread that runs through all these stories. Technological innovations over the last few decades contribute to the sense of being on a treadmill. In comparing their life on the farm to that of their parents and grandparents, many farm couples talk quite philosophically about how technology has brought both positive and negative changes to farming. On the positive side, you can do more faster, with fewer workers, and with less physical strain. On the negative side, you have to be a skilled mechanic to care for equipment, because maintenance is a major task; high debt loads are created to buy the technology and the larger farms that are needed to make it cost-effective; farmers find work more lonely; and the mental strain of juggling debt loads, breakdowns, and the extra tasks of maintenance is high. For example, one couple with a hog operation described the constant concern with finding and keeping good labor. They decided to make a major investment in a new hog parlor, one that would have computerized feeding and gates that would enable the husband to manage the hogs by himself. But this meant taking on a substantial debt to finance the transition, one that seemed much bigger than when they signed the papers because the price of hogs had dropped in the last year.

William Cochrane coined the term *the technological treadmill* many years ago to describe this cycle of adopting new technology, expanding operations to maximize the returns on this investment, confronting narrowing profit margins, and then needing to invest in the next new technology in order to stay in place. But the idea of a treadmill applies to other aspects of farm life besides the effects of new technologies.

When you ask what most concerns them about their farm operations, farm people have a long list. Like most folks, they would like to deal with concerns one day at a time or one concern at a time. Instead, they will work out a solution to one problem, only to have it cause another, or only to find new and unexpected problems springing up before they have fully solved the current one. They encounter a treadmill of continuous problems and constant decisions, few of which work out as planned. It is one of the many mixed blessings of farm life. Farm families value the farm as a place to raise children and to find privacy. But they may not see their children for days when they are in the field, and they cannot miss milking time to take a child to a ball game. We did not find anyone with the magic bullet, the strategy that fends off unwanted consequences in all situations and in all decisions. Like most American families they simply keep living and working, taking each day as it comes and hoping to find a way to emphasize their blessings.

Kentucky's farm families are families like most others; they just happen to live and work in the same place. Although farm families and family farming are wrapped in many myths that bear little resemblance to the reality of farm life, one that does gain substance from our interviews is the importance of family values and the strength of family ties. Although many in society may see the family as an institution in collapse, on America's farms the family remains the touchstone relationship. The farm family is built on traditions, expectations, and shared understandings of their mutual interdependence. This is not to claim that the Waltons is an apt metaphor for contemporary farm families. But it is to assert that the binding stories of family goals, values, and relationships are a powerful force shaping both family life and the family business. We suspect that this would be true regardless of the type of business that the family operated, but the crucible that is farming intensifies the influence of the family. Family ties, family hopes, family dreams, family farm.

2 The Ties That Bind Farming Families

SECOND DAY OF SPRING

When the skies let loose
this morning, I could have, too, but

we were looking for the banker
any time, and there was a hired hand
to feed, and anyhow
someone else might pull up
or the phone ring, or the mailman might
    bleat his airless horn.

I wanted to lay my head on your shoulder
and say, "We work so hard
for so little. Why? How long
can we do this?"
I wanted you to hold me
to let my eyes run
    rain down with the sky, but
there wasn't time and
you never knew
never knew.

BASIN SPRING FARM

I wanted to be a perfect farm wife, and I had a clear idea of what that should be when we moved to Basin Spring Farm in the fall of 1974 with our daughter, not quite two, and our seven-week-old son. I am a girl from northern Delaware who grew up in the country but had never lived on a farm until I was twenty-nine. A girl who spent childhood summers on salt water beaches and sailed in Sunday afternoon races at the Lewes Yacht Club.

My great-grandfather founded a dairy in a small town in upstate New York, where my grandparents and a few elderly relatives still lived when I was a child. Once a year we made the eight-hour drive for a week's visit, which always included a trip to the dairy, run at that time by my father's cousin.

The dairy was always a highlight of the trip. I loved to watch the spar-kling bottles whirl along the conveyer belt, filled in a flash with dazzling white milk, whirl to another machine that capped them, then whirl to a holding area where flying fingers packed them into metal crates for loading onto the delivery trucks. It amazed me that none of the bottles ever shattered or went sailing off the conveyer belt into the room. Our last stop on the tour was to pick up some cottage cheese for my grandmother, the richest, creamiest cottage cheese I had ever eaten, then or since.

The country women whom we visited during those childhood summers were great-aunts in their eighties and nineties. I can still picture their homes and gardens in my mind, but I learned little in those brief visits of what their lives were like from day to day, or how the seasons flowed through the foot-hills of the Allegheny Mountains.

Living on a farm is a far cry from visiting one, so the art of farm wifery was secondhand until we moved to Basin Spring. I learned on the job, un-der the tutelage of my mother-in-law, with my own adaptations gleaned from three years in the Mennonite country of southeastern Pennsylvania.

We had gone to work for Rodale Press, just outside of Allentown, in the fall of 1971. Jim had completed a master's thesis at the University of Ken-tucky that spring that had been waylaid by fourteen months in Vietnam. He had taken a lonely stance, accepting the draft in the spring of 1969, but with his father on the draft board it was one of many stepping stones leading back to the farm, to the sound of the spring gurgling out of the hill and through the basin from which the farm takes its name. That sound ran through the back of his mind in Southeast Asia and helped him focus on staying alive.

The years we spent in Pennsylvania were heady and fun. Jim was hired to manage a new experimental farm and served on the editorial staff of Or-ganic Gardening and Farming magazine. I worked for Anna Rodale and did freelance art work for Fitness for Living magazine. After our daughter was born, she became a covergirl on Prevention magazine and was called on for other photo assignments when a baby was needed.

Sweet years indeed, having babies, working for the magazines, and run-ning the farm at Maxatawny, where we lived in one of three two-hundred-year-old stone farmhouses. It was like being on campus, with friends living and working there, too, and Mennonite neighbors who became our friends and babysitters. We were in the heart of organic America, the Mecca of the back-to-the-land movement. So when we moved from the Lehigh Valley to the Ohio Valley in the fall of 1974, I brought shelves full of Rodale books and magazines and a head full of ideas about how we would inhabit our place in the sun at Basin Spring.

I had enjoyed growing a garden in Pennsylvania and wanted to raise our own food here. My father-in-law plants enough garden for ten families and has always been generous with his abundance—sugar peas, cabbage, okra, sweet corn, potatoes, turnips, cucumbers, watermelons, and eggplant—so I scaled my plantings accordingly. Mine became a "supper garden" of greens and onions, tomatoes and squash and peppers, with my standard zinnias, sunflowers, baby's breath, and marigolds interspersed.

It wasn't long before I graduated from canning tomatoes in a hot water canner to using a pressure canner for green beans, squash, carrots, beets, pickles, and applesauce. There is no feeling of accomplishment quite like seeing the rows of sparkling jars filled with fresh garden produce ready for winter—unless it would be raking and baling twenty acres of hay and loading it into the barn before the onslaught of a summer thunderstorm. But now, with the pressures of my life, I no longer see the gleaming rows of jars. I may no longer have the time to can, but I know I cannot live on a farm without the seasons of a garden. I cannot live without the concentration of the winter planning, the thrill of the first seedlings peeking through the warm spring soil, the satisfaction of the summer and fall harvests, and the simple joys of butterflies, toads, lightning bugs, and all the other critters who come to share dinner with us.

I learned to make sauerkraut using Grandmother's antique shredder, which I found in the pantry of the old house. My baby son and daughter would play in the shade of the old elm in the backyard while I shredded fresh heads of cabbage and packed them into sterilized jars. Sweet corn was always a family affair. The men would pick and shuck the ears while Mary and I brushed the ears, scalded and cooled them, and cut off the kernels that squirted our faces and ran down our arms as we ladled our prize into plastic bags for loading into the freezer.

Twice a day we enjoyed fresh, raw milk, courtesy of my father-in-law, who milked twice a day. And once a week I would churn the cream I had skimmed each day from the gallon jars, pouring it all into the two-gallon Dazey churn that my mother-in-law and I shared. From Mary I learned that a hearty meal washed down with plenty of iced tea was the antidote for a hard morning in the hay field or the tobacco patch. You could never go wrong with pot roast and gravy with carrots, potatoes, and hot biscuits, topped off with chocolate cake or a cobbler for extra energy.

On hot days the children and I would hike to the fields with ice water or lemonade if the men were working in hay or tobacco, a nice break from the canner or the laundry or the garden. I ran errands when needed and kept the farm ledger at night when little ones were asleep.

We were never back-to-the-landers in the trendy sense, but I remained *as faithful as I could to my beliefs about living on a family farm. I was proud of the transformations made by a girl from northern Delaware and proud of my contributions to the family and to the farm at the Basin Spring.*

Farm life changed radically for us, however, in the fall of 1978, when I took a job with the Cooperative Extension Service. It was a hard decision, but one forced on us by the deepening recession in the farm economy. Looking at the books made it clear that the farm could not support the family, even though we thought it should, and so did Jim's dad. It was frustrating, operating the farm the way Jim's parents had, carefully considering each purchase, weighing each decision, scrimping and saving, and yet continuing to fall behind. It was like reaching for the golden ring that was always there but always out of reach.

I have worked for sixteen years, part-time and full-time, for as short a stretch as seven months per job and for more than ten years in the job I now hold. I have experienced the adjustments and compromises of any working mother. The added dimension of farm and extended family makes being a working mother and farm wife more complicated.

Farming is not a hobby at Basin Spring. In 1989 we produced 150 acres of wheat followed by 150 acres of soybeans, 9 acres of tobacco, an acre of peppers, and pasture for between twenty and fifty head of cattle, with additional acres in government set-aside. We rent out our tobacco base and acquire leases, and we share crop expenses with a friend who furnishes labor and machinery, a way of holding down overhead and maintenance. None of this would happen without us, and there are always fences to mend, pastures to mow, and livestock to tend. In spite of all this farm work and production, a question sometimes asked of us by the friends and relations is, "When are you going to farm again?" Indeed. What do they think we are doing today? What would you call 150 acres of soybeans? A garden?

The story I have to tell today is about change, the changes that we and others across Kentucky and rural America have shouldered in order to survive, about what is working in our lives, what isn't, and what we are doing about it. This is not the story that I expected to tell here. This is what has happened. But I believe we are doing what farm families have always done: we are making it work. Somehow, we are still here.

THE TIES THAT BIND FARM FAMILIES

When Barb and Jim came home to Basin Spring Farm they were fulfilling their dream and the hopes of Jim's parents. As Barb notes, she came with a set of expectations formed on visits to family farms in the Allegheny Mountains and refined by their years in Pennsylvania's Mennonite country. During their first years at Basin Spring, the days unfolded to match their hopes and dreams, but then the cold reality of economics intruded. Hard work and sound business decisions were no longer enough to keep the farm business going, and as a result the farm way of life had to change. It is a decision many other farm families have had to make, and to understand the meaning of this change it is necessary to first understand how farm couples define the essential characteristics of their family life on the farm. In other words, when families look at the ledger sheet of farming as a way of life, what goes into the assets and costs columns and how do they balance out?

One asset is the opportunity to be with your spouse, to share your working lives and all that means. It is an opportunity that most married couples do not have because their place of work is some distance from where they live, and rarely do couples work for the same employer. Many of the farm couples spoke with great fondness of the time that they were able to spend together, and it was clear that they saw this as an asset. The opportunities to be together come in many ways. Sometimes it is simply the chance to hop in the truck and go to town together to pick up some feed or machine parts. "Lots of days he'll come through the house and say, 'I've got to run here or there or somewhere else to pick up parts or something else.' I'll ride with him even though I may never get out of the truck. We have a good time as we go and come. We can relax as much in an afternoon doing something like that as some people can on a two-week vacation. It is having the time to talk and be together and do something relaxing and something you enjoy."

More typically, however, the time together involves sharing the work that needs to be done. A farm wife is an integral part of the farm labor force, often because the farm business simply cannot afford the cost of hiring labor. Since many jobs on the farm require at least two people, the business of the farm must involve both husband and wife. For example, setting and stripping tobacco requires the work of several people; it is an activity that couples typically do together. On the farm there are many other jobs that need the help of everyone. A farm wife describes her work: "I will feed the hogs when I know he's not going to be coming in until past dark because he's picking corn or combining. I take care of whatever has to be done. I help cut tobacco, chop it, set it, whatever he needs. I help move equipment from one place to another. I help with the hay when it is busy in the summer, especially if it is the field next to the house. Then the children can play in the backyard and I can watch them."

When you ask farm women to describe the work that they do on the farm, one term is often used: *gofer*. Sometimes the things you need to get a job done are in the barn or the machinery shop, and sometimes they are in town. Wherever they may be, either you have to stop what you are doing and go get what you need or you have to rely on someone else to do the errand for you. This farm wife describes what it means to be a gofer, but her husband interrupts to let us know that from his perspective, her contributions are essential to the smooth operation of the business:

WIFE: In tobacco, I help quite a bit. I set it, chop it, hang it, and help strip. I get out there and help however he needs me.
HUSBAND: I don't think she's telling you all of it. Several times last week it was all she could do to get our lunch to us because she was running back and forth getting us parts when we had machinery break down. It's hard to describe everything she actually does, but she's going all the time. There are so many days that she's as valuable to me as a man or any other person would be and more so, because sometimes a hired hand won't be in a hurry to get me the part and she will.

Farm couples may work together in the fields, move equipment together, and take the harvest to market together. Some of the farm work is done in parallel; that is, couples work on different tasks in different places, but their work is mutually supporting. One type of work that studies show women most likely to perform is keeping the farm accounts. This is a crucial task, and the wife develops an intimate knowledge of what is happening on the farm through the pattern of purchases and sales, even if she spends little or no time in the fields. In many farm families it is the wife who has the overall economic perspective on the farm business, as this woman explains: "I kept all of the books; I paid all of the bills. Now he asks, 'Where is this and where is that?' He has totally left it up to me to do it."

There are several reasons why the farm books seem to be a key job for farm wives. First, they often are the ones to make the purchases of supplies and day-to-day equipment and so have access to receipts. Among older couples, many farm wives have more formal education than their husbands. Third, farm accounting is an activity done in the home and often whenever time is available. It can be done around the wife's other work, such as child care and other household tasks. Furthermore, some farm husbands simply do not like doing the book work—in their view it is boring, repetitive indoor work that seems distant from the "real" work of farming.

As some farm wives pointed out, although book work is essential to developing a total picture of the business and keeping a handle on cash flow, the importance of their role in doing the book work is sometimes ignored. Salespersons and bankers ask their husbands financial questions, when it is really the wives who are most knowledgeable of the farm's financial condition. It is ironic and frustrating, but typical, as thousands of other married women who do their families' book work will attest.

Does the considerable amount of time that farm couples spend working with each other or in mutually supportive tasks affect decision making? Surveys of farm couples about farm decision making suggest that the husband is identified as the one who makes the final decision, regardless of the level of the wife's work on the farm. But this notion overlooks all the conversations that lead to the decision and fails to recognize that the final decision is shaped by an extended process, which includes husbands and wives sharing information about the situation and their choices, discussing the pros and cons of the choices, and each expressing preferences for what should happen. It is true that the process may differ depending on the specific decision being made. Moreover, as this farm husband explains, given the complexity of the economic environment in which farming is done, sharing decisions does have advantages: "I really value my wife's opinions on things and she values mine. I know that if we don't work together I may make a stupid mistake because I will have a lack of perspective on something. That is why it is valuable to me to share the decision making. Then there is the practical side. If you share the decisions, you share the blame. So it cuts both ways. I think that as the decisions get more complicated, we have to work together."

The following exchange illustrates that shared decision making does not occur in all farm families. Some farm women share the farm work but not the farm decisions, although they end up sharing the consequences. There are women who challenge this difference in rights and responsibilities, but others choose not to participate in farm decision making. Their plates are already full, juggling family, home, and off-farm work. The farm is his work, his arena of expertise, and she has her own. Still others take the situation for granted: decisions should be made by the man, and it is the wife's duty to support him no matter what, as this couple explains:

HUSBAND: I am the type of person that likes to do things right and sometimes I go and try to do too much. She has always been there to support me in whatever I've done, even if I've made a bad mistake. One time I made a real bad decision. I bought one farm that I shouldn't have. If it hadn't been for her support and the fact that she didn't try to cram it down my throat that I shouldn't have done this, I don't know what would have happened. Sometimes I talk a decision over with her, but I always make the decision. She supports me. There is so many times in my life as a farmer that has meant so much to me.

WIFE: I feel that this is the wife's place. I may not always agree with what he does, like this one big mistake he made. Okay, he made it. So what? We have to live with that now. There's no need in me saying any more about it. He knows that it was a mistake. Why should I gripe and carry on about it? We're just going to have to make the best of it. And now, with whatever comes from this drought, I feel like that somehow or other we'll come through. We'll be there again for one another.

In the ledger book of farm family life, working together and sharing decisions is an advantage of farming, one not often found in other types of businesses or families. In general there is a mutual recognition that the farm business would not succeed or would not do as well without both contributing their labor and talents. But does this sense of shared responsibilities spill over into other aspects of family life?

When the wife shares the work on the farm, it typically leaves little time for doing household work. And, although farm couples may share the farm work, the house remains the wife's primary responsibility, as it does in most other families. This means that farm women often juggle several jobs on the farm, or else they have to choose whether to do the farm or the household tasks. This is how one farm wife described her everyday life: "The laundry gets done at night. I hang clothes out at night or real early the next morning. Shopping, of course, we're not here to eat very much. I just get the necessities, bread and milk. The house gets straightened up some mornings, and some mornings it doesn't. I figure that if the beds are unmade so what? We're going to crawl right back in them that night and nobody is going to see them during the day. But usually there's some slack time unless I go to the fields that morning. If I'm just running, going and getting what they want, I'll come in and make up the beds or start washing the dishes. They'll holler for something and I'll leave again. I don't worry about it. I do try to keep it halfway straight. But if they can live with it, then I'm happy."

Her comment tells us that for her housework is something that gets squeezed in between the primary responsibilities to the farm. Like many others we interviewed, this farm woman has defined housework as nonessential, or on the list of things we do when everything more important is done. But her solution to juggling multiple work expectations can create some ambivalent feelings. She does "try to keep it halfway straight," but she also "hang[s] the clothes out at night or real early in the morn-

ing." Many other farm women indicated that hanging the laundry out-
doors was one household chore they were reluctant to give up. The fresh
smell of clothes drying evokes a powerful memory of how household tasks
should be managed. It is a part of many stories of what you do if you are a
good farm wife. Another farm wife joked that her neighbors must think
she never did the wash because she usually only had time to hang clothes
out at night and then would take them in early before she began the rest
of her farm work.

A great ambivalence can be seen in these comments: There is pride
in keeping a neat house and doing it as your parents or in-laws did it. But
there is also an understanding that others do not value housework in comparison to other kinds of work you do on the farm. So you feel stress
because you are not living up to expectations, even though you also feel
that those who are important to you—your spouse, for example—devalue
this part of your understanding of what it means to be a good farm wife.
Among most of the farm women who chose to balance housework and
other work by putting housework last, there is a tinge of guilt for leaving
undone, or not doing well, those tasks that have traditionally defined a
woman's world.

Consider, for example, the role of the garden in farm family life. It
rarely is of the size or the importance of earlier years, when gardens of an
acre or two would yield enough food for the cellars to be filled with hundreds of jars of home-grown vegetables. Quite simply, most farm women,
especially those who also have off-farm jobs, cannot afford to take the
time for such an activity, given all of their other responsibilities. And
most farm women would admit that although farm-raised food may taste
better and satisfy a deep desire to be self-supporting, store-bought green
beans cost a lot less than the time and effort it would take to can them at
home. As Barb says, churning your own butter or growing and canning
your own vegetables is immensely satisfying, but the psychic benefits have
to be measured against the costs in time and energy.

In earlier decades a woman's housework was often indistinguishable
from her farm work. Indeed, the farm business depended on her work
because the products of her gardening, canning, and cooking would be
sold to local stores and neighbors or contributed to the in-kind wages
paid hired hands. Many women recall spending most of the day in the
kitchen cooking with their mothers: As soon as they finished feeding lunch
to eight or ten hired hands, they would begin preparing their supper. One
study of Iowa farms found that throughout the first part of this century
hired hands typically received a small amount of cash, housing, meals,
and sometimes clothing as their wages. The work of farm wives provided

much of these "wages." Although hiring labor is more likely to be on a cash basis today, many farm wives still cook and launder for hired labor, especially when a crew of workers is putting in long hours trying to get a crop planted or harvested.

Housework *is* a *chore*—repetitive tasks most members of the family do not value unless they *are not* done. Yet traditionally, the successful completion of household chores has been an important aspect of how women define their role in the family and who they are. When farm women try to balance household and farm work, as well as off-farm work, something has to give; and often it is the housework. Nevertheless, there remains a sense of frustration over not living up to expectations. Many times as we entered a farmhouse for an interview, the wife would apologize for the way the house looked, because "I've been too busy doing" whatever. And there was also some anger that even though she does her share of work on the farm, at the end of the workday when everyone comes inside, the menfolks sit and watch television while she cooks and cleans. As in many families, the family business work is shared, but the housework remains the wife's responsibility.

Decisions about how to manage money matters are often a source of tension for farm families as for others. This common situation is often exacerbated by the difficulty in keeping farm business and family household finances separate. For those who are not part of a small family business, the ups and downs of the employer's business may affect the family if they lead to layoffs, unemployment, or smaller wage increases. But the linkage between the business and the household finances gets played out on a daily basis in most farm families. The family's standard of living is closely tied to the fluctuations of the business, and the economic well-being of the family can be jeopardized by a business crisis because of the intermingling of their finances. This farm wife explains how they try to balance the economic needs of the family and the business: "We live out of the same bank account as we farm out of. A lot of decisions that we make for the family are based on how the farm is doing. So we're flexible enough to know that some of the things that we do are limited because of the farm situation, and that comes first. It's not that the farm is the top priority. We farm to make a living but we also feel like we should have the necessities. But we don't want to put the farm in debt, so we try to keep our priorities straight."

This is not an unusual situation. We met farm families who had a single bank account for the household and farm business needs of two and occasionally three families with joint farm operations. Even when one or both spouses are employed off the farm, the off-farm wages are typically placed into a single farm account from which the living expenses of the family and the business expenses of the farm are drawn. Business and income tax laws discourage commingling, but it is quite common in farm businesses because it does appear to simplify matters. You only have to keep one set of books, and you always know your bottom line. Although nonfarm family businesses rarely commingle their finances, this does not mean they do not struggle over how to allocate financial resources.

It is possible to manage the sometimes competing demands on a single account, but doing so can cause conflict. A decision to buy more land and add to the long-term debt of the farm means that there is less money to make improvements on the home or to put away for the children's education. When funds are commingled, it may become more difficult to accurately assess the economic contribution of the farm operation to family livelihood. This is a situation some farm families found themselves in during the early '80s. The farm had actually been losing money for some time, but the losses were masked by the off-farm wages going into the family's account. Managing household and farm finances in such a situation often means making some tough choices.

One financial choice that farm families are having to make these days concerns whether or not they can afford to have health insurance. The high cost of health care is a problem for farm families, just as it is for every other family. However, since many farm families are self-employed, they do not have the opportunity to purchase the lower-cost group health insurance many other families buy. Farm families must pay the substantially higher price of individual policies. Two couples offer a glimpse of what it costs farm families to carry health insurance in 1988:

"You know Blue Cross and Blue Shield just jumped our insurance rates. We were paying about $216 a month, and now its $360. That's about a 70 percent, and that's just too much of a jump at one time. Health insurance costs are out of hand."

"We were paying $85 a month each. Now it is $152 each. We have to pay $304 per month for just us two. That increase came in the last two or three months."

Health insurance is one of those things that it is better to have and not need than to need and not have. And the cost of having it keeps going up. In the years since these interviews, farm families tell us they are paying nearly two times this amount for health insurance coverage. Farming is one of the three most dangerous occupations in America. The disabling and fatal accident rate is very high. When you add to this the distance from health services and the fact that the average age of America's farmers places them at greater risk for chronic illness, you have a situation in which not carrying health insurance poses a substantial financial risk.

Farm couples have always had to make these kinds of choices when managing their finances. Stories of the early years of rural electrification indicate that the dairy barn was electrified long before the house, and even when the house had electricity, the family waited for the labor-saving devices, such as washing machines, or the luxury items, such as television sets, until after the dairy barn had the electricity-driven equipment it needed to make the cost of electricity worthwhile. To a great extent these were reasonable choices. Other rural families did not have the trappings of the urban lifestyle at that time, and so, when the farm family did without, the differences between how they lived and how nonfarm families around them lived were not so stark. But times have changed.

The mass media and the easy access to urban lifestyles at regional malls have changed the meaning of an adequate standard of living. Doing without, buying only what you can pay for with cash, sacrificing comforts in the home to acquire extras for the farm, and investing in the farm rather than in a child's education are no longer so widely accepted as necessary. Many older farm couples explain these changes in terms of needs and wants: "It was easy to make a living with farming, but people say that now it is hard to make a living farming. What they really mean is that now it is harder to satisfy their wants instead of just their needs. It used to be just people's needs, but now it is what they want. That is why they think it is so hard."

It is the difference between "What do we need to make a living, to have enough to get by on?" and "What is it that we want for a comfortable standard of living?" that seems to make managing farm business and household finances so much more difficult today. Older farm couples comment on this difference frequently, often citing it as the reason younger farm couples have had so many financial and marital problems. Comments such as "They want to start where their parents are now" often began the discussions of how farm family life is changing. But when you think about it, the same is true of most American couples; all of us want for our children more than we have, and certainly more than we started with. To have the trappings of a modern lifestyle—a new car, a Nintendo, designer label clothes, a VCR—is now an important part of how we think about ourselves and judge our achievements, and how others measure our success. In a sense it is the American Dream, and farm families feel they ought to be able to share in it just as much as other American families. A dairy farmer says this explains why one of his neighbors quit farming. "This dairy couple weren't watching their hours, weren't figuring out how much they were making an hour. They worked so hard and had so little to show for it except equity, which could disappear if there was deflation. That was the thing they were frustrated with. They had worked for twelve years and done a good job. They had the top dairy herd in the county, but they didn't have anything to show for it as far as the American dream is concerned. Their house was comfortable, but it was no better than when they moved into it. The material things hadn't improved that much for them. Their kids were growing up and they wondered where the money for college was going to come from. When it is not there now, how can they count on it in ten to fifteen years? This is the kind of young farmers that we are losing because they don't feel like what they are giving to farming is balanced by what they get from farming."

The issue of wants and needs separates the generations in another way. Parents feel that their adult children's wanting and sometimes having so much more than they do is a kind of indictment of them as parents. Does this mean the children feel their parents did not provide for them adequately? Do the children value the material things of a modern lifestyle more than they do the farm heritage the parents have worked so hard to pass on? When the business and household finances are mixed, and the business requires more than it generates, it becomes a problem. Then managing the needs and the wants of the family and the business can be stressful, especially when generations farm together. Farm couples increasingly are wondering why they should be expected to give up on their share of the American dream when others do not. Why, they ask, is it wrong for them to want to have nice furniture or a new car when other families see these as necessities? It is a question that was often asked. It seemed to capture the doubts about farming as a way of life that many couples are now struggling with, far more than their parents ever did.

When you start with different expectations, juggling farm and family finances becomes even more difficult. Historically, young farm women have been more likely to leave farming communities when they reach adulthood than young men. As a consequence, there are fewer and fewer young women willing to marry a young man intent on farming. Over the last ten or fifteen years farm magazines have carried several articles on the increasing number of young farm bachelors searching for mates. In response, some enterprising individuals in the Midwest have started matchmaking services for young farm men.

What has produced this situation? Many an older couple notes that, although they would not oppose their daughter marrying a farmer, they would prefer that she found someone else. Most farm parents agree that farming as a way of life is wonderful; it is farming as a business that makes them less than ecstatic about their daughters marrying farmers. And many of their daughters are not too enthusiastic about the idea either. It was not a universal sentiment among the teenage girls we interviewed, but many stated that the farm life was not for them. Again, it is not just the issue of farming as a way of life, but the economics and the stress of farming that made them interested in another way of life.

Added to this growing sense of discontent with farming and the recognition that it is a tough way to make a living wage for a family is the problem of fewer job opportunities in rural areas. Many leave for college and do not return because their communities do not offer employment opportunities suitable for their skills. Others leave in search of better, perhaps different employment opportunities. One young farmer said that the church his family attends has many older couples and a few younger couples, but there are hardly any young men between the ages of eighteen and thirty. Yet studies indicate that young men raised on a farm who go away to college are far more likely to return to their home communities than their sisters.

As a result there have been more town-farm marriages. Most typically, the town spouse is the wife and the farm spouse is the husband, and they have met either in consolidated high schools or while at college. Town spouses bring with them romantic beliefs about what farm life is like. Barb admits that she came to the farm with quite a few myths about farm life. Part of the struggle is figuring out what part of the myth is true and what is simply a nice story.

For example, town spouses do not bring to the marriage an understanding of the economics of farm life, the nature of farm work, or how closely linked are the farm business and the family life. Nor do they have the skills that would allow them to participate fully in farm work. And although most town spouses are willing to try to learn what farming is all about, the reality of farm family life brings quite a few surprises. One couple describes some of the adjustments that town spouses have to make: "I don't remember what year it was that my husband decided to go into farming with his daddy and they went out and bought a tractor and I almost had a heart attack. He bought a tractor and paid more money for it than we did the house and thirteen acres. I came from a family where my daddy never spent anything unless it was absolutely necessary and he goes out and pays $11,500 for a tractor! I was in cardiac arrest for about three weeks after that. I thought, 'Oh no. We'll never pay it off, never.' I got over that and ever since, it doesn't matter to me what he buys, I have already had my upsets. I have been conditioned."

The economic realities of farm life present the greatest challenge in town-farm marriages; that tractor today could cost $50,000 or more. The size of the debt load, the sheer costs of simply getting a crop in or bringing livestock to market, the long time periods when there is no income coming into the household, and the uncertainty of the yearly income itself all contribute to a sense of frustration and despair, and often marital stress. Several town spouses suggested that although all families struggle to manage their finances, on a farm it is almost impossible to do so because there are so many unknowns.

Another characteristic of farm life that comes as a surprise, and often a disappointment, to town spouses is the pace of life—the constant attention that a farm operation requires and the fact that the weather determines the timing of activities, rather than personal preference. One couple explained it this way: "We didn't get up by the clock or go to bed by the clock. You went to bed by the job. If we had to work until ten P.M. to get a job done, then we'd do it." This is quite a bit different from the way most town spouses were raised. Their parents came home from work every night at about the same time. Weekends and holidays were times spent relaxing with family and friends, and vacations occurred at predictable times. But the pace of life is rarely like this on the farm. One farm husband explained how hard adjusting to the pace of life was for his wife: "This has always been hard for her, and I understand why. I was raised with a farm work ethic. If you weren't working you felt guilty. That was it. You either worked or you were lazy and felt guilty. But her dad was a school principal. While that is not just a nine to five job, he had free time. I felt there was no such thing as free time. When I was growing up, Labor Day meant that was a good day to get a lot done. Holidays meant very little to us on the farm because we would not go on a vacation. I could not take a day off, and if I did, I would make everybody miserable because I was worried about what was not getting done at home. It has been tough on her and it's just that we came from two different kinds of backgrounds."

Finding a mate and the growing number of town-farm marriages are issues that are affecting farm families. There is an awareness that town-farm marriages may carry a burden that makes farm family life even more difficult. Farming families typically raise the next generation of farmers. Growing up on a farm is about learning what it means to be a member of a farm family. When you do not have this background, then you do not have realistic expectations, and you are more vulnerable to disappointments and uncomfortable surprises. For the town spouse, marriage becomes not only an effort to adjust to another person but also a struggle to catch up in terms of knowledge and skills about farming. You are not likely to see yourself as a full partner, nor will anyone else.

For the farm spouse it is a struggle not to impose unrealistic expectations on your mate, but to accept the fact that you may not always share common goals and values. One young bachelor farmer, asked if he thought he would find a young woman who would want to marry a farmer, laughed and said he was not sure. But then he became more serious and talked about the town-farm marriages among his friends: "One of the biggest problems I think is we're seeing more farmers marry city people. You've got one set of values and you've got another set of values. When your values are different it causes problems. It is a lot harder to make the farm work in that case than it would be if you had two with farm backgrounds that understood the farm and knew a lot more about what went on."

His father then offered this final observation on town-farm marriages: "In other words, what he is saying is that a city girl that marries a farmer has a lot of adjusting to make." One farm wife noted there is a struggle to figure out how to fit into a world so different: "You feel like you were visiting from Mars." It may be as simple as accepting the fact that in town, free time may be used to do the little jobs around the house and yard. But the farm is a greedy place; it demands that free time go to taking care of odd jobs around the farm, and maintenance and home improvements can wait, sometimes forever. The common perception is that it is the town wife who is expected to do all the adjusting. She must accommodate to the farm life if the marriage is to survive, not the other way around. She has no choice, if she wants to make the farm and family work.

Money issues always seem to trouble farm families, just as they do many other American families. Through the years, adjustments are made, needs are limited, expectations are scaled back, and wants are postponed. Like their parents before them, some farm couples figure out how to accommodate to the ups and downs in family income. For some, other aspects of the quality of life compensate for these ups and downs. But the pace of life often darkens even these pleasures.

The business of farming is tied to the seasons, the ebb and flow of natural rhythms of birth, planting, growing, and harvesting. The pace of family life is closely linked to these natural rhythms, since they set the time when family activities occur. Older farm couples talk of the early years when life would be hectic for periods of time, and then there would be slack periods when you could catch your breath and catch up on the work you had let slide. It is a rhythm of life that appears in our popular images of farming. The family gathered at the table for a "country break-fast" or evening meal, the family sitting on the porch watching the sun set over fields of shoulder-high corn. Quiet, peaceful, at rest. But the pace of life in the farm business has changed and is changing, and for many couples the change is not for the better. "It used to be you had slack times be-tween seasons. But now everything seems to run together. You're busy all year round. It's more hectic than it used to be."

Many other businesses—manufacturing, retail, service industries—have seasons of hectic activity followed by a time when the pace of work is slower. Workers become accustomed to this ebb and flow of activity; it is a defining factor in the nature of their work life. Farm couples are saying that what has changed is that their lives are all flow and no ebb. "My husband said he thought he'd be finished with the soybeans today, but he won't be. He's worked a full day at least this week moving hogs. He had to take some to market and wean some pigs. Then he had to go by the ASCS office. It's frustrating because some of those are things where you don't see what you've accomplished, but they have to be done. They take a lot of time away from the real work of farming."

As farming has become more industrialized it also has acquired other characteristics of the industrial life. It is not enough just to farm. Today you have to spend a lot of time managing the farm. There is a sense of hurry, hurry, hurry—there are production deadlines, appointments to keep, schedules to meet, papers to fill out. There are increasing demands for management activities, and there is a greater reliance on technology and mechanization. There is a continuous flow of new information about markets, products, and technologies that must be learned, digested, and evaluated in order to survive. These are mixed blessings; life is both easier and harder because of these changes. "It seems to me that besides the farm work, there are so many other activities and pressures to get so much done that you have the feeling of being rushed whether you really are or not. Even though you don't want to think about it that way, you live by a schedule. You've got to get this much done today so you can be ready to do that tomorrow. I think this creates the feeling of being rushed or hur-ried. Sometimes you are not aware of it because you are not punching a time clock like the guy in town does."

When farmers talk about what they like most about farming, they are more likely to talk about the production work, not the management ac-tivities. They like to work at their own pace and make their own decisions on what to do today and what to do tomorrow. But the reality is often quite different. When the hay is cut and lying in the field, you cannot choose to bale it tomorrow, because to get the maximum nutritional benefit from your hay it must be baled with a certain moisture content. Hogs do not farrow to your time schedule, and the dairy cows must be milked on a routine schedule to maximize production. There are established reporting times for participation in government programs. And if your combine breaks down halfway through harvesting the corn crop, you do not have the luxury of saying "I'll fix it tomorrow," for tomorrow may be too late if the rains come, or the killing frost. Increasingly, the freedom of choice and the luxury of setting your own pace are myths that farm couples can talk about only as things they once had that now are gone.

The scale of farming has become larger, and although there are more and bigger machines to do the work, farmers are also producing crops on larger acreages. The machines themselves require maintenance. As the number and types of farm programs have proliferated, the amount of paperwork has also increased and become more complex. A farmer comments on this change in the pace of life: "It seems when I was growing up things were at a slower pace. Now you just stay so far behind. You don't have time for a lot of the things that you did back when I was growing up. Part of it is because you're farming a lot more land, and even though you've also got larger equipment, it doesn't seem to make a difference. For instance, when I was little, we had a pull type combine and maybe forty acres of grain. Now we've got two hundred acres and a self-propelled combine three times as big as the old one. Yet it still takes the same amount of time to get over the ground."

Of course, farming and the farm way of life are simply moving into the twenty-first century with the rest of society. Although some changes in family and work life have occurred gradually over the decades for the rest of society, for farm families most of this transition has occurred since the Korean War. Farm families feel that they have had more changes thrown at them in a shorter time than other businesses. And because farm families do hold more traditional values and have more conservative lifestyles, the changes seem to be more jarring and disruptive.

The pace of farm work is made more frantic in those families with one or more jobs off the farm. The timing of their farm work must fit into the schedule of work set by their off-farm employers. Adding an off-farm job is one way to pay for the costs of the new technologies available to farmers. But it is a choice that also carries costs. These couples juggle multiple work schedules, creating a pace of life that others, looking on, would say is simply crazy. "Well, he had to be at work at 6 A.M. We'd get up and he would go to work. I would go to the hog barn and stay until it was time for me to go to work. I'd get myself ready and during the school year, I always took our daughter to school. We had to work hard. I worked at the bank in Grain Town and had an hour for dinner and my husband got a lunch break too. I would go pick him up and a lot of time we just grabbed us a sandwich and spent our lunch hour in the hog parlor or, at tobacco cutting time, if it was a day that the sun wasn't too hot to burn it, we would go and cut tobacco. It would be ready to hang by the time we got off work. Other times we'd top tobacco during our lunch hour."

The changing pace of the work life has implications for how couples view farming as a way of life. It is not just the fact that the hard work never seems to end, although there are fewer opportunities to step back for a time and relax. It goes deeper than that. It means that there are fewer opportunities to enjoy the simple pleasures that have made farming so enjoyable. The satisfaction of seeing your crops grow is muted by the fact that you are juggling a lot of other work that has to be done and trying to figure out how to do these things and still get your crops in on time. It is this sense of loss, of something precious slipping away, that creates so much confusion and sorrow.

Farming is a greedy occupation. It demands so much, expects so much from everyone who is involved in the business. Many farm families would point to the changing pace of farm life as an example of the greediness of farming. Time is no longer a spendable commodity, for all of it is demanded by the farm. Not surprisingly, many nonfarm couples say their lives are increasingly hectic as well.

Stress is an integral part of living in the twentieth century in a competitive marketplace with a growing sense of economic uncertainty. Similarly, stress has become a part of farming, just as it has for other businesses. Farm families can point to a variety of factors that magnify the normal stresses of life. Key among these is the weather.

Heavy rains or rains that come at the wrong time can be devastating financially. As we learned while we were interviewing, the long, slow death of a drought seems to create an intense sense of stress and despair. Crop farmers have to watch their crops slowly wither and die each day, but livestock farmers must also listen to the toll the drought takes on their animals. "It was hard during the drought. I even quit taking my daily walks from here to the blacktop where our farm begins. I could hear the cattle all over the country bawling because they were so hungry even though they were being fed dry hay. That didn't seem to satisfy them because they are used to that green grass in the summertime. You couldn't walk out in the fields and you certainly couldn't carry hay out to them. They would trample you. You had to haul it with a tractor or stay in the truck. If you got on the ground with a bale of hay they would trample you. If I even started the tractor or the truck, they thought they were going to get something to eat and every one of them would start bawling."

In recessions other business owners must watch their stock sit, unsold, on their shelves, or they must stare at empty dining room tables, and manufacturers must look at filled warehouses and idled machines. In bad times everyone feels the mental stress. But one difference for farm families is that typically they cannot leave the source of their stress; they live there. Every day, twenty-four hours a day, they must watch their business, their livelihood, dry up or wash away or plummet in value, and there is nothing they can do about it. The inability to get away from the source of frustration increases the stress level. One farmer explained that during the drought, even when he went home for dinner, from every window he could see his fields and watch his crops dying. Some farmers move their residence into town, because at least for a few hours every day they can get away from the stresses of their business. As it is, for most farmers, home is not a haven from the stresses of their business life. "The main thing that I think most people don't understand is that when you're farming you never get away from your work. You're there twenty-four hours a day, seven days a week. To me, that is one big drawback to farming. When you leave a city job you go home, and then at least you're away from it. But when you go home and you're still at work and then at night have to watch everything fall apart, you feel you can never get away from it. It seems to suck the life from you."

The phrase *the greedy farm* applies again. As a worker you find that the farm takes more and more of your time. But it also drains you as a manager. Furthermore, there is the sense that you have no control over what is happening. As one farm woman explained: "I don't think people realize the mental stress in farming unless they've grown up with it. As I have grown up and watched my father and then my husband, I've come to realize it's the lack of control in this profession that is so frustrating. With so many other professions, you have more control over the input and output. But you don't in farming. You have a certain amount, but not enough it often seems to make a difference. That's very frustrating."

Others see their situation in more ironic terms. It is the incongruity between what one expects or plans for versus what actually happens that creates their frustration. "When farming, there is nothing that works out like you planned it. You can sit down with a pencil and always figure a profit, and make it look good. But there is always the unexpected that comes along. You can't ever take it into account. It is beyond your control. If nothing else, we will have a drought or too much water. We have made so many great plans throughout the years. But they never worked out. You just have to learn to roll with the tide, I suppose. You either have to learn to live with it or get out."

The stress of farming has consequences for family life. Everyone in the family is affected by what is happening to the business. The common problems of trying to juggle farm and family finances are magnified; and money disagreements that in other times could be worked out become more serious. To survive as a family, you have to learn how to take what comes, learn to accept the things you cannot change, and use the experience to gain more self-confidence in your ability to meet the challenges of farming. "Why do it to begin with? I've asked myself that a lot of times. I would like to be worried about something else instead of worrying about a farm payment or worrying about if it is going to rain. The drought that we had around ten years ago, it seems that we had two or three summers in a row where we got very little water. I was just a nervous wreck the whole time. I kept thinking that if it didn't rain we weren't going to make anything. How were we going to pay for this or how were we going to do that? This time it has not bothered me that much. I guess that I've just gotten used to the fact that you can't change the weather. And even though the drought is worse this time than it was ten years ago, I have adjusted to the fact that you have to take what is dealt to you when you are living on a farm and you are at the mercy of the elements."

One way of handling the stress is with humor. Many farm couples have realized that if they do not laugh at their problems, their problems will bury them. In many of the interviews couples would finish describing how they had lost a crop or had had to borrow money in order to pay the interest on the loans they were already carrying and then tell a joke about farming or farm life. One farmer, commenting on the dramatic swings in the prices that farmers receive and how they make it impossible to ever really make enough money to get ahead, noted the following: "When hogs are high they don't smell so bad. But when they get down cheap like they are now, the manure smells pretty bad!"

Farm couples use many other strategies to manage the stress of their lives. Some are active in softball leagues, others make certain that they leave the farm each year for at least one week, and others simply get in their cars and drive around. Like every other couple who might find themselves at odds, farm couples struggle to find ways to manage the stress and tension in their lives. Most succeed. But there is evidence that a growing number of farm couples fail at this task.

Divorce used to be an urban phenomenon. Farm couples once had the lowest divorce rate of all residential groups. It was once argued that a variety of factors contributed to this low rate: for example, the close working relationships between farm husbands and wives, their dependence on each other for emotional support in hard times, and the presence of relatives to provide financial and psychological assistance led to a low divorce rate. Farm families have also held more traditional values and therefore have defined divorce as an unacceptable choice. It is also true that farm families have recognized the potentially disastrous consequences of divorce for the family business. Today farm couples still have a lower divorce rate than couples living in towns and cities, but divorce is more common. What is happening?

A common perception is that the financial stress of farming has become so great and the work demands so heavy that the bonds that once held farm couples together are no longer strong enough. There is a sense that the continuous pressure of trying to make ends meet, the inability to support the family at a standard of living that is defined as desirable, and the increasing amount of time that farm work demands have become too much for a growing number of farm couples. Moreover, as divorce has become more socially acceptable throughout the larger society, rural and farm people have accepted the view that divorce is an alternative. And so the gap in divorce rates between farm and nonfarm families diminishes. In most of our interviews farm couples could identify at least one divorced farm couple in their community, and sometimes it was a divorce in their own family. "My older son is divorced and has remarried. He has four children. I think a lot of the divorce around here could be attributed to the stress from farming and trying to have a business and trying to make ends meet and never being home with the children."

Another couple told us of what happened to the farmer who sharecropped with them. Their story is about how the greedy nature of farming consumed a couple even when they both came from a farm background. There is a shock when a couple everyone else looked up to begins to fall apart. "The man who farmed our place on share for several years just got separated. He's one of the best farmers around here; he never stops, he works fifteen to eighteen hours a day. His wife never saw him unless she took his meals to the fields. She was lonely, feeling neglected, and with all their other problems—their debt was out of sight—and the constant worry about having enough time and money to get by led her to see a counselor. Now, they're separated and everyone is shocked. They were the kind of farm family everyone around here wanted to be like."

Others, however, point to the growing number of town-farm marriages and the increasing off-farm employment as factors leading to spouses' having different values and expectations. One farm wife commented that many farm women work in town with people whose daily lives are very different from their own. Although the off-farm job reduces the financial worries of the family, it also highlights the taken-for-granted demands of farm life. The realization that other lifestyles are possible has the potential for creating stress in farm families. So when a farm couple works off the farm, they are adding to the personal stress they already carry because of being family members in a farm business. The quickening pace of farm life, differing values and expectations, and the stress of juggling multiple jobs all test the strength of today's farm families. As for couples everywhere, a variety of factors lead to the decision to end a marriage. Living on a farm does not insulate farm families from these pressures.

So what does farming as a way of life really mean? The American story of farm family life is told in terms of serenity, stability, and marital bliss, and we tell the same story about other families in America. There is a kernel of truth here: every couple starts out expecting to have a long and joyful life together. Like every other married couple, farm couples laugh and cry together, argue with each other, stand beside each other in hard times, and grow old together, or sometimes divorce. We heard about many methods used to survive the stresses and frustrations of farm life. A key strategy, as Barb found out, was to compromise on some of your dreams in order to hold onto others. And this requires that everyone work together. "You can get through anything if you really love one another and want to make it work. You may have your disagreements; everybody does. But you have to want to make it work."

A question we asked (and frequently the farm couples themselves raised it before we did) is, Why continue to do it? The answers demonstrate the power of the story of the farm way of life to attract and hold people. Many ruefully noted it was not because they want to make a lot of money. These couples talked about farming as a business more in terms of making a living than getting rich. One farmer put it quite simply, "I don't think you can get ahead in farming. If you don't really love it you'd better get something else."

Many farm couples explained that there are other things in life more important than money. These valued noneconomic aspects of farm life are what farm couples emphasize when they talk about working together, knowing how they depend on each other, or being able to be with their children when they are young. The intermingling of farm, housework, and child care brings couples together in myriad ways. A husband and wife see what each other does every day. There is a mutual understanding of what it means to work and of the meaning of work in the context of a particular way of life. This shared understanding of work is seldom possible for couples who do not work together, and it may be an important factor in the commitment to the farming way of life.

There is something else that helps explain why so many families continue to farm and so many struggle desperately to stay in farming. It is the land. "I think you have to first of all be rooted in the farm as a special place. No one would ever stay on the farm if they didn't love the land." Many commented on the fact that they were farming land that had belonged to their fathers, their grandfathers, or their great-grandfathers. This sense of continuity across the generations was noted as the reason why "I've been trying to hold onto the farm that Daddy had originally," and it explains why many farm families make great personal sacrifices to stay in farming.

The land has meaning to farm families. It is not just the place where they do their business. The importance of the land in how many farm couples balance the scales of farming as a way of life cannot be overestimated. The power of the land is captured in the following comment: "We realize that there is more to farming; there is a culture to the land. I think that is why we stay with it. There is almost a religious quality to working with the land. I'm sure that you've talked to other people who may not have thought consciously about it this way, but have had that feeling of taking care of the land as if it is family."

He was right: not everyone would or could express the importance of family land in shaping their commitment to farming as a way of life. But it appeared in the way they commented on neighbors who did not care for the land properly, who worked the land only for what it could give them in quick profits. And it appeared in their discussions of what it was like growing up on *this* farm with their parents and siblings, the importance of keeping *this* land in the family, even if no family member intended to live on this land in the future. To a great degree family relations and the farming way of life are linked together by the land. One farm wife explained what the farming way of life meant to her in this way: "I did a cross stitch thing that is over our bed. One is for the farmer and one is for the farmer's wife. The one for the farmer's wife says: 'As I chose him I chose this land. I knew that labor is never done. We and the land are one.' We knew what we were getting into when we decided to come home to work his father's farm. We wouldn't trade it for anything."

Sometimes, though, we do not know what we are getting into, and even when we do, it still may be bigger than we can handle. Moreover, knowing and intellectually accepting some of the changes and compromises we have made do not make them any easier to handle emotionally.

3 Growing Up on a Family Farm

WALKING WITH GERARD, WHO IS SIXTEEN

Perhaps a sprite called
"Go with him."
but I changed clothes
    when I got in from work, and
we headed for the creek
to see if the water was up
after the storm.

On the way
we found fish in the pasture
where grandfather's pond overflowed—
bass the color of spring grass
and bluegill shining like pieces of sky
    after the rain. One by one we put them back
    three dozen or more
he caught a bullfrog in his hands
and we admired its legs
before watching it hop back to the pond
and slide into the deep.

He showed me
how he could throw a tomahawk
    into a tree,
and pale jewelweed
    that takes away the itch of poison ivy.

I showed him the site where twenty years ago
his father and I camped with friends
and dreamed
    of building an A-frame cabin,
    a life on the land,
and I showed him the fence row
where we planted walnut trees and loblolly pines
the fall he returned from Vietnam.

The dreams of twenty years washed over us,
youth silhouetted against life,
ideals undone, promises still out.

But we are here
clouds in our hair, cuffs wet with grass
where nine generations have walked and dreamed,
together,
my son and I, our family,
　　　our farm
where the creek sings in May
and fish swim through the grass
into another spring.

BASIN SPRING FARM

My mother-in-law grew up at the foot of Scott Hill Farm in Stith Valley, a twenty-minute ride down the road from Basin Spring Farm, just over the Meade County line. My father-in-law was raised in Breckinridge County within sight of the house where we have lived for over sixteen years. He walked or rode on horseback across Basin Spring on his way to grade school in Irvington. The family moved to Basin Spring in 1929, and he occupied the same bedroom that would later be used by his son and grandson.

Jim has always been a farm boy. He was born in Louisville but was taken home to the farm at Roff on the other side of Hardinsburg. He was four years old when the family moved to Basin Spring Farm, and except for a decade spent in college and the Army and working in Pennsylvania, where our children were born, he has lived at Basin Spring.

One of the first things my father-in-law asked when we met twenty-four years ago was whether I was a "city girl." I did not know how to answer him. Northern Delaware was decidedly rural along the Pennsylvania line during my childhood. The five acres that we moved to in the spring of 1948 on Silverside Road was part of the old Clark farm that had been sliced into six tracts of five to ten acres. There were surrounding farms—the King's, whose loft and spring house we played in frequently, and the Bonsall's, where we would pet the calves and swipe corn from the field to clatter down our neighbors' downspouts on Halloween night. These havens were where my brothers and I and our friends roamed and explored and camped. This personal history, however, never made me a farm girl by Breckinridge County standards, and I knew it. I was as far removed from farm life here as I was from my great-grandfather, Charles Ross Guilford, who milked Holsteins and built the Guilford Dairy in northern New York long before I was born.

Our children have literally grown up at Basin Spring, but their childhoods have been different from their father's, as his differed from his father's. I used to think this was a function of our ability (or lack of it) to make it economically on the farm, and of the compromises that have accompanied our survival. But I have come to understand that the changes we have seen in the last twenty years are mirrored in the broader context of what is happening all over rural Kentucky and rural America. We live in another time, a smaller world than either Jim or I knew growing up in the '40s and '50s. The losses and gains that have put us where we are today are the reason I started on this quest to understand what has been happening to us and our neighbors. I understand now that we are more like other farm families than we are different. But recognizing how we got here may help us in our choices for the future. For, in spite of the twists and turns in the road behind us or the road ahead, I believe that a farm living is good and it can be viable.

Our children's lives have been quite different from those of their aunts and uncles; they have known many of their experiences, thanks to their grandparents. In Tessa's preschool years we were still milking and gardening and making sauerkraut with Grandmother Foote's antique shredder and freezing corn. Their grandparents, Gerard and Mary, were still working with us, doing many of the things that have always been done here. Tessa and Gerard would play around my feet or out in the yard under the trees while I cooked for hay and tobacco crews. Sometimes I worried that I did not devote enough time to playing with them while I was doing something else, like hanging out clothes or weeding the garden. But I took satisfaction in knowing that they were independent little people growing up in a paradise rich in the resources needed to learn and to challenge.

When I went to work off the farm and had to face the issue of child care, Mary and Gerard, their grandparents, stepped in and agreed to take care of the children. Although grandparents have always helped look after their grandchildren, in earlier times it was done occasionally, when needed. But today it means doing it all the time, as more farm parents take jobs in town. For five years my in-laws looked after Tessa and Gerard. I know this gave them great pleasure, but the blessing was offset by the stress of keeping up with two active young ones and the loss of control over their own time. This was not necessarily how they thought they would be spending their later years. But for Mary and Gerard, it was far better for them to take care of their grandchildren than for a stranger to do so.

By the time Tessa would have been old enough to help with dinners for hay or tobacco crews or to can and churn, we no longer did these things on the farm. Times had changed our lives and our farm. For example, mules had been the farm "machines" during my father-in-law's youth, and he has always loved to see a well-matched team. They were used to break ground and to cultivate and were a particularly important asset in the tobacco patch. When Tessa and Gerard were small, we had a team of Belgian horses, half-sisters named Pearl and Cait. We worked them a little and wanted to breed them, but Pearl had foot problems and we sold her. All of us felt that Cait would have made a beautiful brood mare, but we had trouble getting her in foal, and when we succeeded in the second year, we lost the foal before birth to fescue poisoning—a lovely full-term foal with a full blaze. We would have to quarantine Cait from fescue grass if we bred her again, not an easy task on a farm full of fescue. And so, like the chickens, work horses left Basin Spring. After that our primary farm animals were those in our cow and calf operation. But that too has changed.

When Gerard was riding a Honda dirt bike over the fields at age ten, I wondered if he would ever make a farm hand or care about the life that had forged his father and grandfather. But by the time he was thirteen, however, Gerard's interest had turned to horses. First it was the pony left by a cousin who had joined the Navy, and then his grandfather's Tennessee walking horse, a strapping strawberry roan called Strawberry. Tessa and Gerard had ridden him with their grandfather since they were old enough to be held in the saddle. At fourteen Gerard was training a stately black two-year-old named Senator for his uncle Phillip, and the dirt bike was history, to the relief of his grandmother and myself. Besides becoming an expert horseman, Gerard was turning into an all-around hand, fencing and mowing with Jim, hauling hay at Basin Spring and on neighboring farms, and working with the cattle herd. He became a reverent student of nature—one of Jim's dearest aspirations for him—and explored every inch of countryside around Basin Spring, camping in all kinds of weather the year around. He read American literature of wilderness explorations, and studied Western lore and Native American history.

Like most farm children, school years meant 4-H and Future Farmers of America (FFA). Tessa has competed successfully in the FFA creed contest, parliamentary procedure competitions, and livestock judging. In the farm-management competition she has received top honors at local and state levels and a silver medal at the national level. When Gerard reached high school, he followed his sister into FFA and onto the parliamentary procedure and the livestock judging teams and also won a national award in farm management. Although these organizations have meant a lot of time and effort for both the children and ourselves, I have seen the value in terms of their poise and public speaking ability. Participating in 4-H and FFA has been one part of farm life that Tessa and Gerard have shared with those who have gone before them on Basin Spring Farm.

But having children involved in these activities and living on a farm means that you have to take time off from whatever you are doing to transport them to everything kids do today. For us it meant 24 miles round-trip to the high school, 60 miles round-trip to the nearest mall, and 104 miles round-trip for a second medical opinion in Louisville. Getting there is always a major expense, compounded by the insurance costs and worries that come with teenage drivers.

Tessa and Gerard have other things in common with the lives of those children who grew up before them on Basin Spring Farm. They have had the chance to know and be with their grandparents and to play with friends, if arrangements were made. But their lives differ in important ways; they are much less involved in the farm work. Jim put out his first crop when he was fourteen, as was common in generations past. But our son did not have this type of responsibility as a boy, even though he has grown into it. Consolidated schools have given them many town friends whose ideas of weekend or summer fun differ from those of the farm friends Jim had when he was growing up. And they have come to expect different things. Mother Mary couldn't run Jim into town to grab something at the store or to play baseball. But we think nothing about taking the kids to scout meetings or to Wal-Mart. Our children may have to make choices between scout meetings and dance lessons and baseball, but Jim's generation of farm kids didn't expect to have those choices.

Because of the changes in our farm operation that came with our working off the farm, our children have been removed from certain pressures that many farm children experience. In the past and still today, children may leave school at peak work times to house tobacco or pick peppers, and some cannot participate in football or basketball or other after-school activities because they are needed on the farm as soon as they can race home. We have friends who grew up on farms but are dead set against having their children bound to a childhood of hard work. By choice, they live in the country, but not on working farms. On one hand, I can understand this, and I respect it. But understanding is mixed with sadness, for the children miss the valuable lessons a working farm can offer. There is a pride that comes from knowing that your work has helped guarantee the year's harvest, or that your own hands rescued a young calf. You learn responsibility and time management growing up on a working farm.

So our children have not had to leave school for seasonal farm work or devote their after-school hours to the field or the kitchen as did their grandparents. But they have been immersed in the life of Basin Spring in other *ways. They have come to love the farm for what it has provided and what it means. And they have received a rich education from the farm and the community in which they have lived all their lives. While it has not been the farm life I imagined for them when we arrived in a Volkswagen bus in the fall of 1974, fresh from Organic America and the land of the Pennsylvania Dutch, it has been a rich life, and I wouldn't trade it for any other life.*

GROWING UP ON A FAMILY FARM

Farm families are like other families. The joys, the sorrows, the hopes, the dreams, the family fights, the pulling together in times of crisis, the special family celebrations, and the funerals that are the essence of family life on farms are like those that other families experience. Yet in our popular stories of farm life we have built an image of the farm family that makes it seem like something more, something different, especially for the children who grow up on these farms.

Traditionally, farming is a way of life that begins in the cradle. Calvin Beale, a senior demographer for the U.S. Department of Agriculture's Economic Research Service, puts it this way: "It's not easy to recruit commercial farmers from the nonfarm population. Farmers are born, not made—or, I should say, they're made by being born to farm families" (Craig Canine, "A Farewell to Farms," *Harrowsmith Country Life* [May/June 1991]:28). Few nonfarm youths would answer the question, "What do you want to be when you grow up?" with "I want to be a farmer." So in this sense each new generation of farmers tends to be drawn from the pool of children of the last generation of farmers. Being raised on a farm, then, is as much about learning a future career as it is about growing up.

We have always seen the farm as an ideal place to raise children. According to our myths, on the farm children can run and play freely, safe from the problems and threats of the city. On the farm children learn about life and nature and come to appreciate God's works. On the farm children learn responsibility and the value of hard work. On the farm children spend more time with their parents, so the bonds of family are stronger and more complex than in nonfarm families. The description of Gerard and Tessa and the earlier generations growing up on Basin Spring Farm does reflect many of these beliefs about the farm as an ideal place to raise children, as does this comment by one farm father: "You were talking about the advantages that the country children have. I think they grow up quicker. They learn the facts of life being with animals and nature. When our son was just a baby, he went with us when we worked.

We used to tie him in the tractor seat to make sure he didn't fall off. He grew up knowing how to do things. He didn't have to learn how to drive a car. Farm kids just have a lot of advantages."

On the farm children have a freedom to roam and play that many urban children rarely experience. One couple who had lived and worked in town for several years before finally moving to their own farm offered a vivid picture of their children on that first day on the farm. The wife noted that after breakfast she was washing dishes and looked out to see her seven-year-old son running across the hay field, his arms waving and his head up, yelling like a banshee. He never could have done that in town, she said; the neighbors would not have liked it. This sense of safety and the opportunity to run and play at will is expressed by another farm mother: "The kids, they have grown up being farm boys. They're loud. They like to get out and do their wheeling and dealing out in the open. They are pretty independent too. I don't think that I'd feel as safe in town with how they operate. They jump on their bicycles and they go up and down the road. They are so much more safer in my eyes out here, away from all of the traffic and all of the people in town. I certainly don't regret bringing children up on the farm, inconvenient as it is."

The inconvenience this mother mentions comes from the physical distance separating farmsteads. Unlike many European nations, where farm families live in small villages and travel each morning to the family's land, our settlement pattern has been the isolated farmstead. The degree of isolation varies by region. East of the Mississippi, farmsteads are rarely more than a mile or two from each other and typically only a few miles from a town. On the Plains, however, several miles may separate farmsteads, and the nearest town may be more than an hour away. Yet, regardless of the physical distance separating farm families from their neighbors, there remains a psychological distance. The children next door are not a few feet away—over the back fence or across the street—as in many suburbs and towns. The physical and psychological distance is seen by farm parents as both an advantage and a disadvantage.

The physical distance is a kind of barrier to what many see as the threats and dangers of city living. One farm father asserted, "You don't see farm kids getting into trouble like they do in the cities." Although an exaggeration, this comment does reflect the belief that on the farm children do not have the constant pressures that seem to tempt many urban children into trouble. In this sense the isolation of farmsteads acts as a physical and psychological protection from a rapidly changing and confusing world.

Children's freedom and parents' peace of mind, however, does come at some cost. Farm life is fun and free in part because it is isolated. But the isolation can be a problem when children want to participate in after-school activities or weekend athletic events. Farm families are forced either to have one adult spend a considerable amount of time ferrying children to and from town or to consciously limit their children's participation in the things that all the other children are doing. One farm father explained how the coach of the girl's basketball team had stopped by one day and said that if they would allow their daughter to play basketball, he would drive her back and forth. The father said no; if they couldn't do for their children, they wouldn't be "beholden" to anyone else. Afterward, we wondered how the daughter had felt about his decision and her lost opportunity.

Moreover, when you live on a farm, it is not as if you could call up the child next door and invite him or her over for the afternoon. One farm teen put it this way: "The biggest disadvantage that I can think of is not having very many of your friends around that you can ask to come over. You're not that close to them." In Barb's account of Tessa's and Gerard's life on Basin Spring, she does not talk about their having friends over to play frequently, in contrast to her life growing up in a rural/suburban area.

Another consequence of the physical separation of farmsteads is that for most of their early years farm children have each other, their parents, and other relatives as their playmates and role models. It is likely that farm children spend more time with their parents, siblings, and other relatives than urban or small-town children because the farm is both home and business. Many farm parents never really leave the home place to go to work, and for them it is much simpler to bring the children to work with them than to take them to a sitter. Taking the children to a sitter means leaving their place of work, taking time from their workday, and making extra trips, and for many farm couples that is time they simply do not have.

A dairy couple explained how the husband set up a small nursery in the milk parlor so she could take her infant daughter with her when she milked at night while her husband stayed to work the fields. Another mother who has helped her husband work the fields throughout their life said: "As a rule, anytime we're doing anything the boys are usually helping too. Even when they were little, they would be on the tractor. They used to sleep in the tractors and combines. They've always been out with us; they'd take their naps in the fields. When our youngest was nine days old, we had him back there at the silo in the child carrier, and he's been in the truck and tractors ever since."

For many parents, this opportunity to be with their children as they are growing up, to be able to be there when they take their first step or say their first word, is a desirable benefit of farming. Farm couples appreciate it especially when they compare their lives to those of urban friends. One farm wife noted that it had been a factor in the discussions of her husband's finding an off-farm job. "I prefer my husband to stay in farming as his main occupation, because then he has more time to spend with the children. Some of our friends work second or third shift, and their children don't see their fathers except maybe on weekends."

But when you live and work in the same place, child care can become a problem, especially if one or both parents work in town as well as on the farm. Child care is a continuing problem for all working couples. Studies show that couples rely first upon private individuals who take in a few children, next on more formal home-based child care, and then on private, for-profit day care businesses. Although it might seem that child care would be particularly troublesome to isolated farm families, this was not the case among those we interviewed. They rely less on day care centers and more on family and friends, and such informal arrangements seem to meet their needs.

Most farm couples tend to set up their farm operations close to their extended families, and relatives often provide child care when one or both parents also work off the farm. One farm woman said she had left her factory job of seventeen years two years earlier. When we asked why, she explained that her son and daughter-in-law had their first child at that time and her daughter-in-law needed to return to work. We must have looked somewhat puzzled, for she smiled and said, "That's what grandmothers do, take care of the grandbabies." In another farm couple's home the family room had two cribs and a playpen. The wife explained that she kept her sons' children while their mothers worked. She noted she took care of five grandchildren and had made a clubhouse out of the old meat house for the grandchildren. The reliance upon family members for assistance in child care is quite common, and it is not a burden but often a pleasure for grandparents. "We've got three grandchildren, two boys and a girl. They've been here all summer. We show cattle and they like to show cattle. They all own a calf. During the summer we go to all the county fairs around here and they help us show. We get them soon as school is out and we look forward to that day."

In other cases farm couples, like urban parents, rely upon friends and neighbors for child care. However, unlike urban couples, who frequently face the problem of a reliable babysitter's moving to another part of the city or to another city, the residential stability typical of rural communities allows for more permanent and more family-like relationships between caregiver and child to develop. Several farm couples spoke about neighbors who had provided a special kind of day care for their children, sharing in the raising of all the children. It is clear that even when a neighbor rather than a family member provides day care, a close relationship often develops that reinforces the sense of family. One farm wife explained that even after they moved to a new farm several miles distant from the woman who had been watching her son since he was an infant, her babysitter simply followed them. For this farm family, their neighbor who provided child care was more than a babysitter. "She's just like a grandmother to him. He calls her his other mom. She kept him and she would even meet him at the school bus until he got to where she thought he was old enough to stay by himself."

The descriptions of the farm as an ideal place to raise children recounted by the families we interviewed were quite similar, regardless of the individual's age. The farm has always offered a quiet, open space for children to run free, and parents have typically been there with their young children to share in the joys of their growing up. One thing that has changed over the years is the shift toward smaller farm families. As a result the geographic isolation of farm life has tended to cause greater social isolation, especially for younger children, because the number of siblings as playmates has declined. This is one reason why decisions about more organized activities for children have assumed a greater importance within farm families.

But life on the farm is not all fun and games. Farming is a greedy occupation, and children do not escape the demands of farming. There is work to be done, and even the smallest hands can make a contribution to getting that work done. The parents were raised that way, and that is the way they raise their children. Everyone has chores to do, because on a farm the work literally is never done. Livestock must be fed every day; the dairy cows must be milked at least twice a day; fences need fixing, equipment needs maintenance; the garden needs weeding; the barn needs painting; the house needs cleaning. Both Tessa and Gerard have been involved in the work of maintaining Basin Spring Farm, and Gerard's involvement has increased as his interest in the farm and farming has increased.

For many farm parents an essential part of the advantage of growing up on a farm is learning responsibility and the value of work. When you grow up on a farm you have freedom, but the freedom is balanced by responsibilities; there are many opportunities to play and as many opportunities to contribute in important ways to the work of the farm. This emphasis on the value of work and responsibility is a theme that continually runs through parents' comments on why the farm is an ideal place to raise children. "I think a farm is the best place in the world for a child to grow up. They have a lot more advantages. They learn the sense of responsibility for one thing. They don't have a lot of idle time with nothing to do; there's always something for them to do. I've always told them that if they're going to do something, I expect them to do it right. Anytime they start something I expect them to finish it."

Learning the value of work and the lesson of responsibility begins early on the farm. No hands are too small to make a contribution, no job so small that it can be overlooked. Older farm couples, those in their fifties and sixties, recall starting work before they were old enough to understand what it meant to be responsible for a task or to understand the importance of their work for the survival of the farm. "I guess that I started feeding calves when I was four or five years old. That's when I was big enough to pack a bucket and go to the barn. I remember I bought two registered Guernsey calves when I was about eight years old. Chores, feeding the calves was the main thing I'd done for a long time. Then when I was about nine years old, I started driving a tractor by myself."

It may sound surprising for very young children to be doing this kind of work. But nearly all those who grew up in farming told similar stories of beginning work at such an early age. Children's work on the farm continues, and as they grow older their responsibilities expand to match their new skills and knowledge. For many teens, summer is simply the time when after-school chores expand to day-long work. One farm teen noted this difference from his city friends. "I don't know what the definition of bored during the summer is. Other kids talk about a boring summer. I never have that problem. I've always been busy. It hasn't been that bad but there has just always been something to do. You become more independent a little bit younger on the farm. You learn to use your judgment a little quicker."

It is important to understand that children's chores are not just make-work. Rather, children contribute substantially to the economic success of the family farm with their work. As one father explained, once his son was old enough to help, he cut back to only one hired hand; and after his son left to go to college, he had to start hiring more labor. On the family farm every family hand can be a substitute for a hired hand. One mother explains this dependence on family labor: "We've got seven children, four boys and three girls. They had to work. When my husband was at work, the kids would come home from school in the afternoon and we would set tobacco. As soon as they came home from school we'd get a bit to eat and then we'd hit the plant beds and we had it ready to go."

Children's work on the farm contributes to the family income and their own. City children have few options other than babysitting or yard work for earning money before they are sixteen, but the same is not true for farm children. Most farm children receive wages for their work, and it is not an allowance. This may seem like just a difference in words, but it is more than that. A wage is a just compensation for valued work, whereas an allowance is a provision that one makes for someone; it is bestowed, not necessarily earned. "I never got an allowance. It was what I made." Children's income then, comes from the wages they justly earn for making a valued contribution to their family business. "I grew up on a farm and worked on it. I drew a little salary. I did whatever a hand did, mostly tractor work and stuff like that. I had my own beef projects and my hog projects. I was I guess ten when I first started driving the tractor, and my own son started younger than I did. I remember I made fifty cents a day. Dad being the boss, he picked jobs that I could do. I remember the man that owned the farm told Dad one day, 'If he can do a man's work driving a tractor, he can get a man's salary.' That's when I started getting more than fifty cents a day. I wasn't but twelve or fourteen then."

Many farm youths are given the opportunity by their parents to begin their own business activities, typically as outgrowths of 4-H or FFA projects. The organizations 4-H and FFA have long-standing educational programs directed at enhancing the vocational skills and general knowledge of rural youths, but the projects also are a means of earning money. Livestock or crops raised for these projects are sold, and the proceeds are typically the child's. One father explains how his son's 4-H project was the start of the family's cattle operation. "My son was able to take his livestock project and build our herd. He started out with just two cows. We gave him his first two cows instead of buying life insurance for him. I told my wife that instead of a life insurance premium we could buy him two calves and help him feed them. Since that time we have never bought a calf for him. He took those first two calves and grew them and sold them. From the time he has been ten or eleven, I've never used my judgment buying his calves. He would pick them out. We would go to the livestock market; he would pick them out and buy them. If he had one that got sick he would pay his own vet bill. We started our whole herd from that. Now, he buys his own clothes. He bought his own car. But we pay for his school, because we think that is our responsibility."

Although this son's decisions and their effects on his family's farm operation may not be typical, most farm families see children's work as an opportunity for them to earn their own money at an early age. It is assumed that children will use their earnings to acquire their own desired things or to put aside for their future. So farm work provides these opportunities, as well as instilling the value of work and the meaning of responsibility. One young woman explained that she, her sister, and three brothers raised the tobacco, and "the shares were split among the kids. My parents got nothing out of it. They haven't since we've been old enough to work our share. The majority of our earnings was set aside for college. But it was also expected that when we started the next year, the cost of the seed came out of these savings, and if anything else was needed then it was up to us to pay for it from the earnings of the previous year." At nineteen, she figured that she had worked in tobacco for about twelve years. Her comments suggest that parents also use this approach to teach their children about the importance and value of money, a sentiment echoed by many other farm parents.

Yet it is also true that on some farms, the children have few if any work responsibilities. In some cases this is because the parents remember their own childhoods and want their children to have the opportunities for free time and to participate in school activities that they did not have growing up. In other cases the traditions of appropriate work for boys and girls remain strong. Housework is the domain for young farm girls, just as it has been for their mothers, whereas for their brothers, advances in technology have changed the nature of farm work enough that the kinds of chores small hands can do have simply disappeared. So some types of farm work can be done without the help of children, and there are farms where neither daughters nor sons participate in the farm business.

Growing up in farming is learning the value of work and money, accepting responsibility as a contributing member of the family business, and sometimes making the first steps toward your own future career. Working on the family farm also is the way for the next generation of farmers to acquire the skills, commitment, and sense of identity essential to farming as a business. A newspaper story on Kentucky's "Great American Family" states, "The family that farms together stays together," and offers the following comment from the family's patriarch: "I kept them involved in helping me on the farm. Before they were old enough to go to school, I had them driving a tractor" (*Lexington Herald-Leader*, May 17, 1991, C1). Working on the family farm as a child is the way to learn to be a farmer.

Learning the skills and identity of a farmer begins at an early age. Play time often involves doing the things Dad does in the fields. "From a little bitty child up, whatever his daddy was doing in the field, he did in the house on the floor with all his little farm toys. He used to make hay bales out of paper, and the rows on my big braided rug was his field rows."

For farm children, especially boys, the key memory of growing up on the farm is the first time they drove the tractor. Driving the tractor serves as a rite of passage. It marks the transition from doing chores, what one farm youth called "menial" work that simply has to be done but is not important, to doing a "man's" work. All of the adult men and the younger farm males that we interviewed could vividly recall the first time they drove the tractor and really helped their dads. A teenage son describes that first time: "The first day I drove the new combine home, I was little. It was a 6600 John Deere; it was a big $50,000 to $100,000 machine. I came from the back field by the big pond. My dad had been working back there when he had a problem with the tractor. He came back to the house to get what he needed to fix the tractor and took me back with him in the truck. When he was done, he didn't want to have to drive the truck back home, and then walk to the back of the farm to get the tractor. He put me up there and told me all I had to do was leave it in forward. Just go forward and pull back to stop and steer. I wasn't very old, just a little bitty dude. When we came home my mom said all she could see was this little blond head sticking up driving the tractor over the hill. From then on I loved tractors."

There is one other important aspect of passing on the farming tradition. It is the growing awareness of the meaning of the land. Family farming is typically bound to a place. Growing up in farming is as much coming to understand and appreciate the land as it is learning the skills and knowledge. This is especially true when the family has been working the same land for several generations. "I remember Dad talking about what the farm was like when he bought it. I remember the day that we bought the additional land and walking over it with Mom. They were so proud of it when they got it paid for. It meant a lot to them and it means a lot to me. I'm trying to give that feeling to my kids. I feel like they are proud of what we've done. I think they'll want to keep it."

Learning the skills of farming—how to make the land bountiful and how to raise productive animals—is part of what it means to grow up in a farm family. It is learning how to recognize when something is wrong with a cow near calving or how to identify the first signs of blight in a field of corn. It is learning to hear when the engine on the tractor is running rough and knowing when a heifer is freshening. And for many farm youths it is coming to understand and appreciate the land as something more than a place where the family business is done. On some farms it is recognizing that a particular stand of trees has been the site of childish games for generations, or realizing that a stone fence was laid with loving care by a great-grandfather, or that the land has fed and cared for the family for decades. And it is learning about the capacity of the land to produce successful crops: what fields can't be worked till late spring and what field has a problem with Johnson grass. This is knowledge that is indigenous to a particular parcel of land and is learned in companionship with those who have worked the land before you. But as Barb explains, farm life today is different for her children than it was for her husband, and these changes have consequences.

As the nature of farming has changed, so have the social expectations about raising children. Although farm children have always been actively involved in the work of the family business, many of the older couples pointed to some important differences between the way it was when they grew up and the way it is today. Some of these difference can be attributed simply to changing expectations and values, but others can be traced to the increasing technical complexity of farming.

Many farms have become more specialized, and so the diverse set of tasks that once provided many opportunities for young children to do chores year-round have disappeared. As Barb notes, there are no more chickens to feed, and the milk cows and the work animals are gone as well. Now, many children help only at peak times, when the tobacco is being set, the hay is being brought in, or the cattle are being taken to market. Bigger, more complex machinery and the increasing use of chemicals have meant more complex tasks that require more maturity and size to do. There is a sense that farming may not be enough, that children need to develop other skills that will enable them to find good off-farm jobs, just as many of their parents have done.

And some are coming to feel that the future of the family farm may not be with their sons, but their daughters. In the past, women tended to become farmers as the result of a family transition—the death of a father or a spouse—and they operated the farm while waiting for a son to grow up or until remarriage. But today some young women are making the conscious decision to pursue farming as a career. Their brothers may have decided that farming is not for them, or they may have no brothers. We interviewed younger women who were operating the family farm because farming was the life they wanted; it was the career they had chosen. We spoke also to older women who did not think their sons would choose farming as a career, but they themselves desired to continue the work they had grown to love.

Another difference is the meaning of children's work within the context of family life. The fact that farm children typically earn wages for their work appears to be a more recent phenomenon. The oldest farm couples talked about their work on the farm as children as a family responsibility. Work was something that everyone in the family was expected to do, a family duty or obligation. Children did not expect to receive wages for their work, nor did parents believe that they owed their children wages. What parents gave, frequently, was a share in the income that came from the work. Just as there is a difference today in many farm youths' minds between an allowance and a wage, there is a difference for older farm couples between the shares they received and the wages their children expect. The meaning of children's work, according to many older farm couples, has changed from a familial responsibility to a kind of market exchange. Although they agree that this may be more fair for the children, especially since some of the older farmers felt that their parents had not treated them fairly, there is also a feeling that something has been lost in this transition. The family solidarity that comes from shared responsibilities, mutual obligations, seems diminished by the emphasis on wages.

Another difference between children's farm chores today and yesterday is their potential contribution to a child's economic success in later life. Farm chores may no longer be the stepping stone to acquiring skills that will have economic benefits as adults. The pace of technological change and the transformations in the scientific knowledge base of agriculture often make skills learned as a youth obsolete. The days of having children quit school to work on the farm are fast disappearing. Parents are more likely to stress the importance of staying in school than permitting or encouraging teens to leave and help on the farm. The relative balance between a good education and the teens' work on the farm is now tipped toward school, not farm work. "There was work to be done when the kids came home from school, whether it was to strip tobacco or whatever. I'd go out and have everything ready for us to go to work. A lot of times we'd strip tobacco late into the night. If the kids had homework they usually didn't work in the fields. But there were times when the kids were in grade school and we'd have trouble getting the tobacco in. I'd run down here and get them and we'd take down the tobacco and then I'd take them back to school."

It might be surprising to hear a parent say, these days, that children would take time off from school to do farm work, but in the past this frequently happened on the farm. You do what you must to survive. If the tobacco is cut and still in the fields when the rains come, you could lose half of your year's income. It is a difficult choice, but one that farm families have always made. Many older farm couples never completed high school because their parents needed their labor to keep the farm and the family going. Today, it may be as simple as choosing between letting your son play in the regional baseball championships or getting the hay in before the rains come; or it may be as serious as keeping the children in school and not getting the hay in or keeping them out for a day or two and being able to make your mortgage payment this year.

Increasingly, this choice is not available to farm families. Schooling has changed; it is like a full-time job. The structure and momentum of learning in schools is such that children can not afford to lose several days to farm work, because they will fall too far behind. Moreover, the tolerance of local school officials for this kind of absence is nearly gone, as funding depends upon average daily attendance and state education officials establish higher standards of accountability. Other school activities, such as clubs and athletics, also have become more important in providing children with skills that will be useful later in life or can provide them access to financial assistance for college. For both Tessa and Gerard, their involvement in FFA and 4-H has meant opportunities to build skills and knowledge that have led to college scholarships and will someday lead to other types of personal success. Hence, the potential economic gains from staying in school and participating in school activities may now be equal to or even exceed any contribution that farm tasks can make to the child's career performance as an adult.

One aspect of having children live and work on the farm that has not changed over the years is the potential hazards that await young children on the farm. Farming is a dangerous occupation, even for adults, and for children, who often can be easily distracted, the farm and farm work can be life-threatening. Tractor rollovers and equipment with many moving parts able to trap and injure are obvious potential hazards. But even the most innocent situations can be a problem. "My parents' first child drowned when she was eighteen months old. She got away and got in the pond that was behind the house. My parents were pretty protective of us after that. My dad wouldn't let us do anything that he thought we might get hurt at. I don't remember ever riding on a tractor until my husband and I were going together. My dad never would let us do anything like that."

Clearly, the technological revolution has changed the kinds of tasks that children can do and the age at which they begin working. Larger, more complex machinery requires more maturity and sometimes simply greater size on the part of the operator to be used effectively and safely. And some tasks, such as applying herbicides or pesticides, are rarely seen as safe for children to do. Farm parents recognize that these changes make it more difficult and dangerous for children to work on the farm. "We have two boys, sixteen and thirteen years old. They're a lot of help. They started driving tractors around twelve. They didn't start as soon as their daddy did. But all of the equipment is so much bigger and more complicated. I started on an old Farmall with an eight-foot disk, but it is kind of hard for a little kid to run an 86/40 with a thirty-foot disk."

The technological nature of farming today, the large and complex machinery, the increasing reliance on chemicals, and bioengineered inputs all contribute to making farming one of the most dangerous occupations in America. Take a trip through any farming community, and you are likely to see teens and adults who have lost limbs; or read any rural newspaper, and you will see a report of a child or an adult injured or killed in a farming accident. In the last few years the issue of injury prevention on farms, especially for children, has become a topic of national debate. For health and safety professionals, the following story by an adult farmer is an injury or death waiting to happen: "When I was little, I had a little dirt spot out here in the corner of the yard where I played with my tractor. I was probably about five or six when I first drove a tractor. I was so small I couldn't even reach the pedals. I had to stand up to drive it because I couldn't reach the pedals sitting down. I'd get it going and then I'd hop up in the seat and steer."

There are many types of hidden hazards in the urban or small-town home, but there is a difference in the nature of the farm hazards and the time required to obtain appropriate treatment. Besides the large machinery with many moving parts, the chemicals, and the farm animals, there are farm ponds that tempt children in the heat of summer and work to be done in the depths of winter. If something happens, it may be that you are in a back field, half a mile from the nearest phone, which may be half an hour or so from the nearest medical help. These conditions may account for the high accident and mortality rates for farm adults and children. The hazards are real, and although children may see only the pleasures and the fun, their parents never forget that the farm is not always benign.

The responsibility for work and for their own earnings teaches farm children a valuable lesson. Their involvement in the labor aspects of the family business often carries over into the parents' willingness to bring the children into the financial affairs and decisions of the business. "We always have discussed in front of our kids all the major decisions we make, like buying the land that we are interested in. They know when there's no money in the bank. They know when payments are coming due and that kind of thing. We just didn't feel like it was good to exclude them from the day-to-day operation of the house, farm, or anything else. They need to know what we are doing. They know what is going on, they know what the financial situation is and that kind of thing." Indeed, there are many times when young men are actively involved in the routine management of the farm. The modern farm business requires a broad range of technical knowledge, and young people acquire it in 4-H, in FFA, and in vocational agriculture classes. Some parents come to recognize the value of this knowledge and encourage their children to take an active role in the day-to-day affairs of the farm or to assume greater control over their own incipient business enterprises. In other cases, parents have helped a child start a 4-H or FFA project, and the project has grown into a major part of the family's operation.

The years of growing up on the farm, learning the skills and knowledge of a farmer and sharing in the closeness of a family that depends upon everyone's contributions to survive, culminate in a decision either to go into farming or to leave. But farm youths today are not choosing farming as a career in the same numbers as their parents did.

In earlier years the choice was simpler, for there were few alternatives to farming. Jobs outside of agriculture were very few in rural communities, and many farm youths simply did not want to leave for other types of jobs in distant cities. Moreover, the labor requirements of farming led many parents to exert pressure on their sons to stay and work the farm with them. And if a father had died, there was a prevailing community expectation that at least the oldest son would stay and work the farm for his mother. With few other options nearby, with an awareness of how much the family farm needed their labor, and with a community sentiment toward staying on the farm, there were strong pressures to choose farming as a career.

However, the situation is very different today. Although completing high school was often not seen as necessary to be a successful farmer in the past, today it is. The decision to go into farming is sometimes now made following high school, when a youth decides whether to stay and work the farm with his family or to go to college. Others are encouraged by their parents to go to college, because they recognize that increasingly, success in farming depends upon acquiring all the technical and business-management knowledge that is available. "Everything has to be so efficient, there's such a fine line of profit. There is not any room for mistakes. I would say that if anybody is thinking of farming now, if at all possible they should have a four-year college degree in agriculture. Then they would have all of the farm management they need and everything else added to it. But how are they going to get the money to get that degree? And once they get it, they will probably go into something else. They wouldn't want to farm!"

Here is the irony of the situation. Although young adults considering farming as a career must go to college to get the education and technical skills necessary to have a chance to make a go of it, that process may actually jeopardize their going back to the farm. William Heffernan, a rural sociologist at the University of Missouri, has been surveying the attitudes and ambitions of agricultural majors for more than ten years. He recently commented: "The thing that has surprised us in the last couple of years is the deep anger and resentment that many of these students, who are mostly farm kids, express about the long hours, great risk and small financial returns involved in farming. They're bitter, and they're not going back." Not only are these farm youths not going back, but many others are not going back either. "Of the 12 national officers of the Future Farmers of America in 1986, not one was planning a career in production agriculture" (Canine, "Farewell to Farms," 31-33).

What is happening? There are two characteristics of the current situation that make it substantively different from earlier times. First, many parents are not encouraging their children to follow a career in farming. Indeed, many are actively discouraging their children from this choice. For example, an older couple explained that none of their children had returned to the farm. The father, with a smile, noted that when his daughter quit college to work the farm with them, he had given her "the absolute worst" jobs to be done, and she went back to school that fall. A young couple who were struggling to make it on their own farm said that when they had told his parents they wanted to farm, they asked them why would the couple "waste" their college education farming? The couple then said that his parents had offered only minimal assistance in setting up their farm operation. There is a sense of resignation and determination that underlies many parents' views on their children going into farming.

Parents want their children to understand the differences between farming and its alternatives so they can make a knowledgeable choice. The following answer to the questions, "Would you want your child to farm? What would you tell him about farming?" indicates that parents are quite aware of the differences between farming and off-farm jobs, especially when they've tried both: "I'd tell them that they'd probably be better off if they didn't. If they really want to do something else, they would be better off to do something else besides farming. Like I said, if they want to farm, I'll try to do all I can to help them farm. But as for a preference, I've seen both sides and they'd probably be better off not farming.

Farming was about the only choice that I had. It is a hard way of life. We've talked to our kids about better ways of living. There is farm life and then there's city life. There is a difference in going to work at eight in the morning and quitting at four in the afternoon, or getting up at four in the morning here and starting your work and then working until everything is finished. And doing this with no certainty at all of any fixed income, no security, no insurance, or any of these things."

For many young people, the years of working alongside their parents, of sharing the joys of a good crop and the tragedy of drought, and of seeing the debt load get bigger because the prices of livestock seem to be the same year after year have taught them some bitter lessons. One is that despite hard work some people never seem to get ahead. Another is that a farmer's work is never done. One farm youth described the desire to have a "normal" life: "A few of us talked about how we liked farm life but we wanted to have a social life or be able to get up and say well, I'm going to go shopping or I'm going to go to the theater or do something besides get up and go hoe out tobacco or run cattle or stuff like that. We want to get up and have a nine-to-five job. Farming is getting up at two in the morning because the cattle are out or getting up at six in the morning to start off your day and not coming in until nine or ten at night. So for a lot of us, it's the long work hours, and we just want to have a normal life."

Moreover, farm youths simply have to look around their communities to see another reason for thinking seriously about a career other than farming. There is a near absence of young (ages twenty-five to forty-five) farm couples. The average age of farmers today is fifty-two, and it is estimated that there are now twice as many farmers over the age of sixty as there are under the age of thirty-five. Farming is no longer a career chosen by many young persons, nor is it a career that seems kind to those starting out.

There simply have been fewer young persons entering farming in the last few decades, but more significant was the farm crisis of the early 1980s, which had devastating effects on the generation of farmers who had entered the business in the '70s and early '80s. Studies of farmers in North Dakota, Iowa, and Texas who either went bankrupt or voluntarily left farming to avoid losing all their assets revealed some disturbing points. More than half were young farmers (ages twenty-five to forty-four), nearly half had college degrees, and most had at least two children. In effect, we lost a substantial proportion of the next generation of farmers. The following is an apt description of what is happening on many farm roads in rural America. "Me and one of the neighbors was talking about two weeks ago. He is about the same age I am and he said that in ten or twelve years he'll be quitting. He said, 'You'll be quitting too.' And he said that others would be quitting because they were older than we are. 'I don't know what is going to become of things. I don't know who is going to be farming then.' I'm sort of like him, I don't know what's going to happen. Out to this end of the road there isn't but one that is younger than I am. He is just about a year behind me. The rest of them is all older. I'll soon be sixty-two. All of the rest of them are older than I am and are getting up around seventy."

A look around their communities would raise serious questions in any young person's mind about the prospects for getting into and succeeding at farming. The events of the early '80s served as a warning that farming is a career whose financial risks may well exceed any family or lifestyle advantages. The young farmers who went out of business during the crisis of the '80s were not bad managers or too eager for financial success. Many were recognized in their communities as the brightest, hardest working, and most technically knowledgeable. The young farmers starting out in the '70s followed in their parents' footsteps, making the same types of business decisions that had brought economic success to their parents. The problem, as one social scientist noted, was that the times were different; the decisions that had led to prosperity in the '50s and '60s were disastrous in the '70s and '80s. For young adults considering a future career, the message seemed clear: Farming is not only a gamble, which many would be willing to take, but it is also a game in which the rules are always changing, and the players never know when or how they have changed. Quite simply, many have decided that there have to be easier ways to make a living. "Many kids have already tasted working for money. They aren't going to work for nothing. That would be the hardest thing for somebody that has worked for a regular salary and is used to having money to do. To start to farm and maybe wait for three years for a paycheck and live on borrowed money—I don't think it would be possible. They would never learn to do without the things that they are used to having, to make a sacrifice they would have to make to live on a farm. They wouldn't be willing to work and put every penny back in like we've had to do, and still not see a profit sometimes."

So what are the assets and costs in the ledger book of farming and farm life from the perspective of growing up in farming? For children growing up on the farm, life is playing, running free, learning the value of hard work, having an opportunity to gain some financial independence, and being with your family. But, as in all families, there is a balancing act.

Sometimes family members must make compromises between what they want to do and what needs to be done. Parents recognize the potential hazards of having their children work on the farm and the real need to have that extra pair of hands to help set the crops when the spring rains have delayed planting or to get the tractor back to the barn when the storm is coming and the truck with this year's corn also must get under cover. And helping Mom and Dad on the farm does not seem like fun or a learning experience to a child or teen who really wants to play with some friends.

Sometimes the memories of the hard work and the family togetherness are bittersweet. A young farmer offered this answer when we asked whether he resented having to do farm work all the time when he was growing up. "I did then, but now I look back I can say it never hurt me as an individual. I only went twice to the public pool that I can remember. I hardly ever went anywhere in the summer. People would say, 'What do you do all summer? Where have you been?' I'd say 'I've been working.' I regretted it then, but now I don't at all. It kept me out of trouble, that's for sure. It did do that."

There is a sense for some farm youths that living on a farm far from town that needed everybody's help meant that they missed something that other children had. In the preceding comment, you hear the sense of loss about not being able to go to the pool in the summer. It is one echoed by many other farm children. Just as driving the tractor is a rite of passage to adulthood, spending time at the swimming pool in the summer is the celebration of youth—one that many farm children express regrets over missing. There are responsibilities and more adult perspectives you must assume at an early age when you grow up on a farm. There are things like work responsibilities, financial responsibilities, and accepting the fact that the farm comes before what you want to do. Sometimes, then, there is a sense of loss that leads to some bitterness about a youth that resembled adulthood more than childhood. "When I was little I hated the farm, and I still do, because I was never allowed to do anything unless it revolved around the farm. I couldn't participate in a lot of school activities because I had to come home and help with whatever needed to be done on the farm. I mean in summers I couldn't participate in softball because we had to cut tobacco or we had to go pull cane out of the bean field or something. It makes you resentful sometimes, living on a farm, because you think, if I lived in the city then I could be playing sports, or I could go out on a date if I wanted to. But living on a farm it was, I can't do this because my mom would say it'll take too long to get into town or to pick you up. Living in a rural area is real nice: it's spacious and it's not as cramped, but you're also deprived of a lot of stuff that kids in the city have."

There is both the bitter and the sweet in these memories. The resentment did not often come out as openly as this, but sometimes you could hear it in the wistful comments about vacations or recreational opportunities missed. Living in the country, even play can take on a different meaning when it's something that *has* to be done. "We got to go bike riding, but mom always found creative things to put bike riding with. My mom would say, 'We need this, ride your bike to the store and pick it up for me.' It's like mom, that's five miles away. And she'd say, 'You can do it, it's good exercise!'" Fun and games are not always fun when they become necessary. Several noted that when they were little they would play by helping their parents hoe out gardens or tobacco patches, but that the "fun" wore thin when this activity changed from a game to a chore.

The changes in the pace of farm life and the growing number of parents with one or more jobs off the farm also mean that parents have less time to spend with their children than in the past. For many farm families the only time they sit down to eat together these days is on a rainy weekend, when Dad cannot be out in the fields trying to catch up on farm work he could not do during the week because of his job in town. Or perhaps the only time parents see their children is when they are sleeping. If parents are concerned about the loss of time with each other because of having to juggle too many tasks, their children are feeling this loss too.

Finally, both farm adults and farm youths have come to recognize that the physical isolation of living on a farm does not always insulate children from the social problems of the larger society. Parents may assert, "You don't hear about farm kids getting into trouble," but it is said more as a hope than as a statement of fact. In several interviews with parents and children, there were matter-of-fact discussions on where the drug parties had been held that weekend and where teenagers could go for alcohol. Drug use, vandalism, petty thefts, and more serious crimes are growing in rural communities. But often the physical separation of farmsteads hides these problems; we simply do not see our neighbors' children getting into trouble. So the freedom that comes from being able to ride your bike anywhere and yell without bothering the neighbors can also be an opportunity to get into trouble with no one to observe. In a troubled world, living on a farm gives no certainty of immunity.

Thus, a tradition of generations seems to be faltering. The straight road that once led from being born on a farm to learning the skills and lore of a farmer to making the choice of farming as a career is now interrupted by attractive alternatives and in some cases simply blocked by the economic realities of the agricultural marketplace. Some farm youths continue to live on the farm but do not farm. They make their living in the factories and businesses that now populate rural communities, or they commute to sprawling suburbs for even better employment opportunities.

Fewer and fewer farmers are now being made by being born to farm families. As a result the age structure of farm communities is beginning to look like an inverted pyramid. Young couples still choose farming as a career—there are just fewer of them. And to a certain extent those who do are exhibiting a level of commitment, a faith in the future, and a willingness to assume a financial burden not expected of young persons entering other careers. Although neither Tessa nor Gerard has firmly stated, "This is what I want," the possibility is there. When young people today choose to accept the torch from the previous generation, they are looking for a harvest of hope.

4 Merging Families and Business

BASIN SPRING FARM

Four generations of the Foote family have lived at Basin Spring Farm. My father-in-law dreamed of owning this farm when he walked across it as a boy on his way to school in Irvington. His father owned it first and moved the family into the old frame house in 1929 when my father-in-law was in high school.

Grandmother and Grandaddy had lived here nineteen years when they sold it to Jim's father and mother in 1948. Jim was four years old when the family moved to Basin Spring, and he grew up here with two brothers and two sisters. The family had lived at Basin Spring for twenty-nine years when we bought the farm from Jim's parents in partnership with Jim's brother in 1977. Eleven years later we bought out our partner, and since the spring of 1988 we have been on our own.

The house stands as a legacy to generations and gives full definition to the old adage "It takes a heap of livin' to make a house a home."

Indeed this is true. The bedroom our son occupies is the one used by his father when he was growing up and by his grandfather when he moved here as a student in high school. Our daughter's bedroom, always known as "the girl's room," is the only upstairs bedroom with some heat via a vent in the floor to the dining room below. It has been occupied during the girlhoods of two aunts and several great-aunts, one of whom died here at the age of sixteen of appendicitis. Two of our nephews, born in the room where I am writing, farm today with their father in Henderson County. And a grandfather eight generations back, who owned two thousand acres in the area that encompasses Basin Spring Farm, is buried on the adjoining land that Jim and I deeded to his parents when they sold us the farm in 1977. Multigenerational has been the evolution of Basin Spring Farm.

For Jim, it began in childhood. He was only four when the family moved to the farm from Hardinsburg. Whether the bug bit him then or after, I do not know. But as long as I have known him, Jim has always wanted to farm.

At fourteen he put out his first crop and with the profits bought the John Deere 530. The first weekend I spent at Basin Spring Farm was in July of 1966, and I had to meet the 530 along with the rest of the family. I re-

member walking to the tobacco barn, where Jim started the engine so that I
could experience the singular "chooca-chooca-chooca" sound that only a
two-cylinder John Deere tractor can make. But it was more than the voice of
the 530 that I heard that summer morning; it was the sound of a legacy.
Years of labor and love were contained in that "chooca-chooca-chooca." I
do not know exactly when I knew we would come back here to farm, but
the seeds of that understanding surely date back as far as that July morning.

Our son, as soon as he could toddle, climbed up on the 1530 John
Deere that we bought our first spring here. If he has dreams of farming today, they have more to do with horses than with cattle and wheat and soybeans and tobacco. But I know something is stirring within him when he
saddles his horse and heads out across the fields, where he disappears for
hours at a time. Whether his imaginings will bring him back here I do not
know. Or maybe it will be our daughter, whose farm-management team won
a silver medal at the national convention of Future Farmers of America. The
seasons will tell.

Jim worked on the farm the summer and fall after he graduated from
high school. When he enrolled at the University of Kentucky in January of
1963, he took the advice of both his older brothers: "Major in something
that you like," they had said. They had gone to college to prepare for careers in industry and did not share Jim's desire to return to the farm. So he
studied agriculture and eventually earned a master's degree in agronomy.
We moved back to Basin Spring Farm in the fall of 1974, and our life as the
third and fourth generation on Basin Spring Farm began.

There was some wonder in the family and in the community as to why
we would want to do such a thing. From more than one person we heard
variations of "With your education, why would you want to farm?" or "Why
do you want to farm if you can do something else?" "Why indeed?" our
banker echoed. During our years in Pennsylvania we had made a farm payment plus interest every six months on thirty acres adjoining Basin Spring
Farm that we had bought when Jim returned from Vietnam. But our banker
balked when it came to putting up operating capital for a full-time farming
operation. So we found another banker outside the county.

From mid-October to the first of December 1974, we lived with Jim's
folks in the old house while we worked out the financial details that would
get us started. Jim's mother fixed a bedroom upstairs into which we moved
a Mennonite cradle for our son and a daybed for our daughter plus all the
plants I had packed into our Volkswagen bus. She had washed and ironed
the tieback muslin curtains and put a braided rug by the bed. To warm the
room, my father-in-law laid wood ready to light in an ornate little drum stove

that had belonged to Jim's grandmother. That completed our cozy quarters, until we moved into our own home three weeks before Christmas.

That home was a fourteen-by-seventy-foot trailer near a grove of trees overlooking the spring. Tessa called it the "new house," and it was pleasant, with morning and afternoon sun, a view of the pond, and trees that dropped their leaves with a delightful tickle on the roof in the fall. Year around, stepping outside the trailer, you were greeted by the murmur of the spring that was the source of the farm name. It was this sound that ran through Jim's mind in Vietnam and fueled his determination to return from the war, back to Kentucky. To come home to Basin Spring.

That first winter was pleasant, the holidays stretching into January with visiting to and from the main house and laying groundwork for the coming spring. It was an exciting time of change and planning. In February we added another John Deere to the repertoire, the 1530, with a front-end loader that proved to be very functional. It was a bulldog of a machine, and we thought we had virtually signed our lives away buying it. But that meant that Jim and his dad could work together.

We broke ground in February that had not been plowed for twenty years. As the work began to evolve, Jim's mother and I took turns cooking for the hands who were hired on a seasonal basis for such major jobs as putting up hay or tobacco. Sometimes our nephews came down from Louisville to help during peak times, but most of the time it was the four of us working here together, making it all work.

A friend of mine describes her experience with the sweet side of a multigeneration operation that includes her brother-in-law and his wife and her husband's parents. The men would plant the garden and her mother-in-law would tend it, she said. "Mom would call and say she was doing beets or corn or beans, and I'd help her all day long." The two women would visit on the porch while they were snapping beans and talk about everything under the sun, but often about the family. She, like me, relished the other "simple little pleasures," like walking down to watch the men mending a fence, or stopping to visit them in the shop when they were working on the farm machinery during the "slow season" of the farm year, or hanging a fresh wash out on the line.

Ours was an insular life compared to life in Lexington or Pennsylvania. At times it was extremely lonely for me, as I knew no one outside the family. But there were great rewards such as being around cousins and aunts and uncles on a regular basis, and as my friend described, I became totally caught up in the seasonal flow of farm living.

Jim and his father worked well together, though there was never a formal partnership agreement between us. However, it was not the same rela-

tionship that they had enjoyed when Jim was fourteen and putting out his first crop. At age thirty he was no longer "the boy," which caused some disillusionment on both sides, I believe. But for the most part the work unfolded smoothly.

My role as a daughter-in-law was very subordinate—to the farm and to the family, in that order. Coming from a nonfarm background, I do not see how I could have proceeded any differently at the time. I came to understand early on and in the most graphic terms that there was a world of difference between visiting a farm and making a home and livelihood there. I had a lot to learn and I knew it, but I immersed myself wholly and with enthusiasm. Over time I learned to stake out a life within the limits of family and economics. We had to survive, but I had to make it work for me too.

We farmed with the folks for three years until our rather different agendas called for new paths. We needed to be gearing up economically, and the folks were gearing down. But our years together gave us a foundation on which to build. The children grew to know their grandparents at work and at play, whether we were planting wheat or hauling hay, freezing corn and gathering tomatoes, or fishing in one of several well-stocked ponds. When the farm changed hands, the folks moved into a new house on twenty of the thirty acres that Jim and I owned adjoining Basin Spring Farm, and we gladly moved into the roomy old house.

Our partnership with Jim's brother was also distinguished by different agendas. He lived in New Jersey for four years, then Louisville, while we lived and worked on and off the farm. Many discussions of what we were selling or buying took place in long distance phone calls. After eleven years it seemed that our investment and his might be better served if one of us bought the other out. And another step in our evolution began. Jim and I went out on a limb again in the spring of 1988, borrowing the money to buy out our partner, as we purchased Basin Spring Farm at market value for the second time.

Some of the feelings of a multigeneration farm still linger here, even after the changes in management and ownership. Our enterprise spans four generations, and our home embodies much history. My brother-in-law still comes down to ride horseback with our son. The rest of the family drops in regularly throughout the year. And at Christmas time we cut an eleven-foot cedar tree, decorate the house, fire up the wood stoves, and open up the big old rooms to as many generations—upwards of fifty people—as we can crowd together over spiced cider, fruitcake, country ham, and whatever people care to bring to the gathering. Our nieces and nephews are beginning to introduce their young children to Basin Spring. The farm and the house continue to be the touchstone for generations.

WHEN GENERATIONS FARM TOGETHER

Basin Spring Farm is like hundreds of thousands of other farms across the nation. It is a place of business, it is a home, but most importantly, it is the hearthstone that ties together several generations. This chapter examines how farm families set up a multifamily farm operation and work out both family and business relationships.

The typical image of the multigeneration farm is one of a father and son working together on jointly owned land, sharing the work and profits, but this is only one way in which a multigeneration operation can be formed. What is common among the many forms of the multigeneration farm is that there is some type of link to the family. The oft heard joke illustrates this point: "How do you get into farming? Inherit it or marry it." But the real question is, What are you marrying or inheriting?

The answer is often "the land." For many farm families, maintaining a working tie to the land is important. The land is more than an input to a business enterprise. It symbolizes a family's history, their relationship to the physical and spiritual world around them. Basin Spring Farm is the land where several generations of the Foote family have made a living and a life. Particular physical locations have special meaning—the spring that gives the farm its name and has provided water to those who have lived there, the place where a great-grandfather is buried, a field where a husband planted his first crop. This sense of identity with the land is echoed by many other farm families: "We all farm together. I bought the farm that I was born on three years ago and my son owns half interest in it. It's just eighty-two acres, but there's a lot of history there, and now my son is planning to live there."

The commitment to the land often is expressed in farming practices. Farming the same land generation after generation gives a special knowledge of the land and its productive potential. You come to know what grows best where, which part of the farm you can get into first in the spring, and which part tends to pool water in heavy rains. Farmers will sometimes choose not to participate in government programs if they believe the program requirements will reduce the productivity of the land. Others make costly investments to reclaim land that has been overused or abused. One farmer explained that he had bought some adjoining land that had been heavily cropped for several years. The land had lost almost all of its topsoil. It had gullies that you could drive a pickup into and lose the truck, he said; the erosion was that bad. It took ten years, but every year he would set aside some money to bulldoze an eroded area and reseed and fertilize it. He rebuilt the land. With pride he explained that now it was good pasture land, although it would never be able to be cropped again. Others explain their relationship to the land in spiritual terms: "You take care of the land and it will take care of you. It is part of my responsibility in farming to take care of the land."

For these families and many others, having a child or other relative to take over the operation upon their retirement is crucially important; and whether consciously or not, at least one of the children is encouraged in a variety of ways to consider choosing farming as a vocation. But there are also mixed feelings about this matter. Farm parents are very adamant about not wanting to "force" a child into farming, because they know all too well the heartbreaks as well as the joys of farm life. Yet there is also a deep hope that a child will decide to farm, so that the family farm—the family land—will stay in family hands. What is difficult and painful is the realization by parents that they may well be the last of the family to farm a place; that none of their children want to or will come back to farming. There is real pain in the voices of the farm couples who have come to see that this hope will never be fulfilled. "I was raised right here. This is the reason we will probably never sell this farm. It has been in our family five or six generations. I wish one of the kids had wanted to stay and farm, but no one wanted it. I don't know if the kids will ever be able to divide it up or do anything with it after we are gone, even though none of them will ever be likely to farm it."

For this father, keeping the farm, the land, in the same family has great importance; it continues a long tradition. It is also clear that although his children know this, he recognizes that he may well be the last of the family to farm the land. Yet this does not mean that the farm will pass into new hands. Rather, the father hopes that somehow the children will retain ownership of the family's land. Maybe for someone in the next generation? One farmer in his sixties was continuing to operate the family farm even though he wanted to retire. When asked why, he noted that although none of his children were interested in farming, his grandson (age ten) spent every summer with him and often talked about wanting to be just like his granddad. He said, somewhat wistfully, "Maybe he'll farm the place someday."

This desire is one reason why farm communities are dotted with idle farmlands or tenant-operated enterprises. Although the children have no desire to farm, they are also reluctant to sell the family land. What they have inherited is a physical memory book of family history, and you simply do not sell your family history.

At a more practical level the multigeneration farm is also desirable because it is a family business. The emphasis, in other words, shifts from the farm to the family. There is a desire and an opportunity to maintain and elaborate on family relationships throughout one's life. It is the fact that Jim and his father could work together and the children could get to know their grandparents and other relatives in the context of both work and play. The theme of family ties appears in the assessments of many other farm families. "I set up a corporation and I gave my three sons stock in it; they own probably 51 percent of the corporation between the three of them. They get a salary and then a thousand bushels of beans apiece a year for their shares in the corporation. Actually, I could hire help and probably come out with more money myself than having a family unit but that's not what it's all about and not the way I want it."

Farming is a business, but it is a family business. So sometimes you make decisions that are in the best interest of the family but not necessarily of the bottom line. This family owns and operates several farms. Each son has his own farm and jointly operates his own and the parental farm with the other members of the family. It may not be the most profitable arrangement for the father's operation, but that is not the point. The purpose is to have a family farm. Another young farmer who had started his adult life with an off-farm job explained how his desire to enter farming with his father led him to some important decisions: "I had turned down two or three job offers to manage some other peoples' farms, some pretty big farms. I turned down those jobs to work for Daddy in a family farming operation. You have a lot of security being in a family operation. You feel like you take care of each other. You're not just an employee of someone."

The importance of the family and of nurturing family relationships weaves through many descriptions of the multigenerational farm enterprise. However, besides the emotional support and security found in working with family and the value of maintaining a family tradition, the family farm has other advantages. Associated with the family farm is a reservoir of cheap family labor, the detailed and historic understandings of the natural resources associated with the place (such as soil quality, runoff patterns, weed problems), and perhaps most importantly, usually lower costs to enter the business.

Why do most people say that to get into farming you have to marry it or inherit? It is because few people have a half-million dollars or more, the amount the U.S. Department of Agriculture estimates is necessary to start up a farm operation that is capable of supporting a family (what is called a commercial-level operation). What it means is that if a young couple decide that they want to enter farming, they must have assistance from someone. The ways in which families work out this assistance typically reflect the financial condition of the family, the number of children who might need assistance, views on family rights and responsibilities, and the nature of parent-child relations.

The ways families assist children, siblings, or other relatives to enter farming are myriad and run the gamut from outright inheritance, free and clear, to simply offering advice. Several farm couples describe the various ways families can provide assistance. First there is inheritance free and clear of any encumbrances. Most commonly, however, that occurs when there is only one child and therefore the parents do not have to deal with the issue of how to treat other children when the farm represents the substance of the inheritance. In the situation described below, the inheriting child has worked for many years on the family farm, providing evidence of an interest in farming and "building credit" in the enterprise. His father describes the inheritance arrangement: "I'm sixty-six. Financially, I don't need the income off of this farm. My son is the only child that we have, and when he gets through college in two years he wants to farm. I've told him that he can have everything he makes off of the farm. I'll help him work the farm. He won't have to hire any labor, where I've one hired hand now. The machinery, the dryers and everything, I'm getting it absolutely in A-1 shape, so when he steps in two years from now, he shouldn't be out any expense on machinery or anything else for four or five years. That puts him in a different situation than a kid going out here and trying to get started buying some used equipment that breaks down and having to pay interest on borrowed money. I set him up a trust fund when he was ten years old. It is a pretty big trust fund. In fact, if he farmed for two years and didn't make a dime his trust fund would take care of it, at least for two years."

Far more typical than this kind of inheritance arrangement is the tendency to bring a child into an operation by providing opportunities that one normally would offer only to relatives. For example, parents may allow a child to "buy into the farm" with labor. Sometimes parents will pay wages, but in other cases the child receives a portion of the farm products as payment for the work. The child may use the income from the sale of these commodities to purchase machinery, as Jim did on Basin Spring Farm, or put it toward the purchase price with parents. Or the income may be used to rent or purchase a small farm that the child operates alone or adds to the family's productive acreage. These approaches are illustrated by the following descriptions of how families got into farming:

Buying in by sharing the cost of equipment: "When I was in high school, I had a half interest in a hay rake and a baler that I used to custom bale hay. Then, when I graduated, Daddy asked me if I wanted to farm all the time. I told him that I did. So I just started buying in on the equipment with my Daddy. He helped me out a whole lot getting started because he didn't charge me interest."

Expanding the parent's farm operation through purchase of additional land: "All during my time in college, I was farming actively with my father. When I got near graduation, I began taking over a little bit more of the operation. I rented some land and expanded the grain operation in addition to working with Daddy's dairy. Then, we bought our own farm. When we first got married we got paid four hundred dollars a month from the milk check. Of course, the grain crop, Dad just more or less turned the grain over to me and I paid the expenses and I'd get the income. I started buying machinery and taking over all the repairs on the other machinery. So basically Dad just kind of turned over everything to me and the income was mine to pay the bills with."

Multiple siblings entering into the family operation: "I wanted to farm all along. But Daddy said he wasn't in the shape where I could come and farm with him when I got out of the Army. So I took a job in town. Then, after a couple of years Daddy said it was time and I could come to farm with him and my brothers. My brothers were working farms they had bought with Daddy's help. So Daddy and I bought a small farm together. It was about eighty acres. Then I swapped my interest in that farm for that many acres of the home farm and bought the rest of the home farm from him. Then we bought another farm together. Now all of us operate about a thousand acres. We each own our own farms but we swap equipment and help."

Sharing work and equipment with relatives: "To start with we rented a tobacco crop from my grandfather, and then we rented his place. It just sort of all worked from there. Every year I added a little bit more land. Right now Dad and me kinda farm together. We are in on the major pieces of equipment, but we each have our own of the smaller equipment. When it comes to planting corn and beans or combining, we work together. But he owns his land and we own ours. Then there is one farm that we rent together and each gets a portion of the income from it."

From many farm couples, you have a sense that helping children to get into farming is a parental responsibility, although children are expected to "work" for this assistance. Clearly, the ways in which the family assists the child will depend upon their resources. Sometimes all the parents can afford to offer is their labor or access to their machinery. But, as many of these farmers noted, this is as valuable as financial assistance. For some families this reciprocal arrangement of parental responsibilities and children's obligations is formalized through the "hiring" of the child's labor and the payment of an hourly wage. Or the parents may agree to cosign a loan that enables the adult children to purchase land and begin their own operation. Yet even when parents speak in these terms, it is clear that along the way children have been provided opportunities to begin building equity in their future, as indicated in the following comment. "I started my son out on an hourly wage where Daddy didn't do that with me. If you work then you get paid. Now I'd like to start him out in some hogs and see him accumulate some money and then maybe he can buy a farm. We might have to sign a note just like my daddy did. But there's no way that you can start out without somebody helping you. Now you don't have to give it to them; they can earn it. Maybe he'll do some work for the use of the machinery or something else that lets us work things out that way."

It is clear that these relationships are mutually beneficial. The children have provided labor for the farm, but in exchange, besides their wages, they have an expectation that if they choose to go into farming, they will receive help from their families. If the adult children go to college before returning to the farm, the family farm will also benefit from the new knowledge and skills they bring back.

An interesting aspect of these arrangements is how farm families differ on the issue of paying wages. Many older farm couples noted that when they were growing up, children worked on the farm for the collective good, not for financial remuneration. The value of a child's labor was its contribution to the success of the enterprise, and the return for this labor was the profitability of the family farm. The emphasis was on the family welfare and the economic progress of the family; individual goals were subordinate to family goals. In contrast, younger farm couples are more likely to make explicit the monetary value of a child's labor contributions by setting an hourly wage, thereby affirming the importance of the child's labor contributions to the family farm. The wages often are reinvested in the family operation, at least for a time. But with an hourly wage, children have greater freedom to exercise decision making, to make their own choices about the use of their funds or what to do with their lives. It is no longer assumed that a child will remain in farming and therefore eventually receive a return from invested labor. Rather, an hourly wage may be the parent's way of indicating to a child that vocational choices are open.

What we have seen is that farm families are exceptionally creative in assisting children who are interested in going into farming, reflecting a combination of factors such as the size of the original family farm, the parent's economic situation, and the preferences of the individuals involved. However, there are some common approaches that reflect as much the nature of family farm life as they do conscious efforts to assist children in entering farming. For example, nearly all of the farm couples recalled having their own livestock or other projects that provided their own income at an early age. Typically, such income was invested in strengthening their ability to enter farming independently (for example, buying machinery, livestock, or land) or reinvested in the family farm in anticipation of eventually taking it over. Also, most farm couples could point to specific examples of exchanges of labor or equipment with parents or siblings once they had begun farming on their own. Both of these approaches illustrate long-standing traditions on family farms. Children typically have had their own economic activities, and families and neighbors typically have exchanged labor. These activities once had (and may still have) sound economic reasons for their existence, but they are also important as ways of assisting children in establishing their own family farms and helping them to save for college so they can bring more skills and knowledge to the farming enterprise.

Finally, parental assistance to young persons starting out in farming is nearly universal. Although changes in the economy and other social trends may no longer lead farm youths to see returning to the farm as *their* obligation, parents usually see a familial responsibility to help as much as possible. As their parents did before them, and as they will do for their children, many of the assets of the family business are gathered and saved and then shared with the next generation. In some cases this may mean helping the child-partners get into farming, even though it may not necessarily be on the original family farm.

Whatever method families use to bring children into the family farms, all must eventually confront the problem of managing business and family relationships. What makes American agriculture somewhat unique is that it typically is a *family* farm, a business with several families functioning in the same physical space. Such cooperation can be very problematic. In the next sections we examine how multifamily farm operations work out business and family relationships.

Basin Spring Farm may be the homesite of four generations of the Foote family, but it is also the economic enterprise that allowed these Foote generations to have a roof over their heads, food to eat, and resources to educate and care for their children. As Barb explains, though, at various points in time the families involved in the Basin Spring Farm came to

view the business in different ways, and these differences had consequences
for the family farm operation. Working out business relationships in multifamily farm operations can be the source of conflict or of strengthened
family relationships. The outcome depends on how families balance business and family relationships and decisions. What decisions need to be
made and what relationships resolved?

Perhaps one of the most important issues to be resolved is whether to
enter into a formal business relationship. Most farm operations have a
net worth in the hundreds of thousands of dollars, if not more, and commonly thousands of dollars are spent and earned each growing season.
Multifamily farm partnerships have economic production activities on
common land and shared risks in terms of labor, financial inputs, and
returns. Not all such partnerships involve formal or legal partnership agreements. Whether a multifamily farm business develops a formal partnership agreement can be a source of great tension among the families. The
issue touches upon the nature of parent-child relationships, questions of
inheritance, and individual expectations.

Many who described how they got into farming implied that "it just
sort of all worked from there." One farmer called it a "sorry arrangement"
because he did not really know what he was getting paid; it was simply
what his father gave him. Such a situation is not unusual; in fact it is how
many families get into a partnership with their parents.

Q: How did you handle the partnership with your dad? Did you talk
about it?
A: No, he won't talk about it. That's hush hush. We just went into partnership. We just started splitting the money. I didn't buy into anything.
For a long time, we'd simply split the milk check. I took over half of the
debt when I went in. The farm is still in my dad's name. The partnership
return is filed for income tax purposes, fifty-fifty tax advantage. But there
aren't any corporation by-laws or anything like that. It is more or less a
verbal type of thing. It is a partnership right now just for tax purposes.
Our portion of the salary is a little more than it used to be. It used to be
that we got the same salary as my dad. But now that I have taken more
responsibility in the milking, our portion of the salary is a little more
than what my dad takes out.

Why are formal agreements so difficult to negotiate? A formal partnership agreement legally defines the nature of the parent-child relationship and specifies what happens to jointly owned property in the event
of death or the dissolution of the partnership. A formal agreement requires all parties to address directly difficult personal issues (for example,

the death of a parent, divorce, and inheritance) and indirectly consider others (for example, how decisions will be made and the allocation of resources among the households). Children are reluctant to force parents to clearly define economic relationships, and parents are reluctant to confront their own physical frailties and mortality.

A formal agreement is a contract, a public legal arrangement typically between strangers. It is not always viewed as appropriate for private family matters. Indeed, some would argue that a formal partnership is a corruption of the family relationship, which should be based on "He's my dad and I know he'll do what's best for me," rather than "I need to be sure that everyone knows what's his and what's mine and how we will manage the business." Formal partnership agreements seem to imply suspicion, a lack of trust, and the need to have the law protect you; they are often viewed as negatively as some would view a formal prenuptial agreement.

Furthermore, some people feel a formal partnership can place parents' assets at risk. Although the divorce rate remains quite low among farm families, many of those we interviewed could point to multifamily operations that were economically disrupted or were actually forced to sell off a portion of their assets during the property settlement phase of a divorce. Typically, this is a concern when there is a "town-farm" marriage; that is, the daughter-in-law is not from a farm background and so is seen as not having the same commitment to farm life or the same values as other family members. In these situations daughters-in-law are viewed with some suspicion; they have to "prove" their commitment to the farm life.

The occurrence of divorce in farm families may increase this suspicion and widen the gulf between the new daughter-in-law and her husband's family. The reality is that in divorce the wife is more likely to take assets from the family farm business, because typically it is the woman who has married the inheriting farm son. This situation also happens in nonfarm family businesses and, aware of the legal and financial implications of these situations, farm parents feel they are justified in being concerned about how a divorce may affect the family's farm business.

Many farm parents gradually ease their children into a partnership arrangement. As a teenage son or daughter begins to add his or her own livestock or equipment or land to the family's operation, he or she begins to act as a partner. Yet few of us would establish a formal partnership agreement with a teenager. So parents and teens just continue working together, and the time to formalize the relationship seems to slip by. It is often easier to simply continue doing as you have always done than to struggle with the difficult decisions necessary to formalize an agreement.

But frustrations seem inevitable when there is no formal agreement. Adult children often see it as a way for their parents to continue treating them as children rather than as adult partners.

Finally, a formal partnership agreement becomes particularly problematic when a couple has several children, not all of whom want to farm, for such an agreement ultimately has consequences for the children's inheritances. Without formal agreements, the child partners cannot be fully assured that upon the parents' deaths their investment in the joint operation will be accorded its proper importance in the division of the estate. Conflicts may develop among siblings because informal understandings between parents and the child involved in the operation were never made explicit to the rest of the family. Or the informal partnership may lead the child partner to make some assumptions about the division of the estate that the parents have not intended. The emotional turmoil that accompanies the inheritance decision when formal agreements between parents and child partners have not been made is clearly illustrated in the following accounts. The first involves a widow with three sons who is struggling with how to set up the inheritance, and the second is an older farmer's recounting of events in the late 1940s.

"The division of the estate is a problem when you have three sons. How do you do that? My older son already had his farm when my husband died. If I had been eighty years old then, things would have been quite different. But I wasn't ready to retire at forty-nine. I was too young for Social Security. I had to do something, so my youngest son and I kept the business running. Since my oldest son already had his own farm, more than likely the farm will go to my younger son, or at least part of it. It could be divided up between the older one and the younger one. I really don't think at this point that the middle son would want part of the farm. But you never know. There is hard feelings there. I am not looking forward to trying to decide how to divide it up."

"My dad had a right smart of land, oh, probably four hundred acres, when he died. I was in the service and thought it was my duty to come back and try to take over running the farm. But it didn't work out all that well. My mother left this farm and her other property to my sister and her husband, even though I had stayed and worked the farm and taken care of my mother. It hurt me. I felt like I had been more or less disinherited. After my mother died, all I wanted to do was pay my sister that five thousand dollars for her share of the farm estate and never have to deal with her again."

For this farmer, the memories of what happened to the family farm following his mother's death are still bitter, and they have shaped how he will deal with his own children. In partnership with his sons, sharing some common land and machinery, he has set up a formal corporate arrangement. He is working to insure that his sons are aware of the terms of the will, and the family corporation clarifies the economic relationships among the members. As he explains, "I'm trying to lay things out. It's all going to be three ways, and it's one thing I want to do right bad. Anything I do I talk to them about it. It may make them mad, but at least they know what I am planning and what will happen. I'm not going to have the same thing happen to them as happened to me."

Formal agreements can provide both economic certainty and security for the families involved in multifamily farm businesses, yet not without some costs. Parents feel that demands for formal agreements question their fairness in the treatment of their children and raise the specter of their own mortality. Children feel that without such an agreement their investment in the joint operation may not be properly acknowledged in the eventual division of the estate. One farmer summarized the business problems inherent in a formal agreement in this way. "If they could sit down and put it on paper or make a contract everything would be all right. But then you get mistrust there too. You think, well my son is trying to take advantage of me or something. But if you could somehow write a contract that says over the years this will be turned over to you, or somehow or other work out a plan to gradually turn the farm over to the son, that would be best. Yet at the same time, the father needs to get something out of the arrangement too, and that's what makes this such a hard situation. Often the father doesn't want to charge the son what's a fair market price for that farm. But at the same time, he's worked all of his life on that farm and he simply can't walk away with nothing."

Although this discussion has focused on father-son partnerships, multifamily farm operation can also be by siblings. Sibling partnerships have some similarities to father-son partnerships but also differ in important ways. For example, there is an assumption of authority in a father-son partnership, but this is less true when siblings are partners. And sibling partnerships are more likely to involve absentee partners, who have received a share in the farm as part of their parents' estate. Absentee partners may look at the farm business differently and value different outcomes than the partner who is living and working on the farm. The following illustrates some of the differences that may emerge in a sibling partnership involving an absentee partner. "My brother and I were in partnership for about six years. I had started out in partnership with my father, but then he decided he wanted to retire. He would only let me buy out his share if I went into partnership with my brother. I thought my brother's involvement would give us leverage in expanding and improving the farm. We put in equal investments up front. It was my job to run the farm and make it pay its own way. But we hit a couple of bad years, and the farm lost money. My wife and I had to look for work off the farm to pay our bills, and my brother took the write-off on the farm losses. We wanted to invest in improving this place and make it grow and be more profitable; he wanted to get the most return for little or no investment. It became clear that we saw the farm differently, had different reasons for being involved with the farm. The time had come to end the partnership. We struggled to figure out how to end the partnership in a way that was best for everyone, given that he was family. But we couldn't ignore the differences any more."

Not all sibling partnerships end with the farm partner's buying out the other siblings, but maintaining the family business with absentee partners requires reaching understandings about goals and expectations. Absentee partners must accept the economic realities of farming as a business, and on-farm partners must be able to explain the possible consequences of decisions, both those made and those not made.

The farm business confronts a host of decisions every day. Some of them are short-term production decisions, such as when to plant or harvest, what type or amount of fertilizer or herbicide to use, or when to move a herd from one field to another. These are the decisions that keep the business running on a day-to-day basis; they involve the division of labor among family members as well as the expenditures of small amounts of cash. Although short-term decisions, they are crucial to the economic success of the operation. If you apply the wrong amount of herbicides, you may harvest as many weeds as beans, or you may lose an entire crop of tobacco to blue mold. Some partners work out short-term decisions by establishing arenas of authority based on experience or interest and fall into a comfortable arrangement of work and decision making that allows the business to function fairly smoothly. "We talk between ourselves about what we're going to plant and how much. Certain decisions are kind of split up. I take care of the chemicals and fertilizers, and I've got all of the ASCS [Agricultural Stabilization and Crop Service] paper work pawned off on me. My dad does all of the cattle buying. Then, as far as the selling of the crop or the cattle, we very seldom sell anything in any volume without talking it over. We raise seed beans and seed wheat. I usually end up taking care of most of that. I get the fields worked out as to what we're going to plant where. He takes care of leasing all the tobacco. Then, as far as the actual farm work goes, we each have our own separate jobs that we do."

Sometimes the decisions and the allocation of work activities reflect what each partner is most comfortable with and knows the most about. An advantage of a multifamily farm business is that among the partners there typically is a broad enough range of knowledge and skills to fill all the gaps. It also means that an older partner may not need to take the trouble to learn how to use new equipment or new chemicals. Partnerships, then, relieve some of the burden of operating a farm in an ever-changing environment. "My dad is sixty-two now and still will decide on what crops he puts out, but he leaves the decisions on what herbicides and insecticides to use to me. There are some pieces of equipment, especially the newer and more complicated things, he won't use. He doesn't really care about having to learn how to use them."

Implicit in these kinds of work and decision-making arrangements is a mutual respect for the partners' skills and knowledge. In each of these cases the father and the son recognize that each has something valuable to contribute to the business operation, and each one is willing to allow the other to assume some control over that aspect of the business. But not all family businesses can come to these terms, even on the simplest decisions. One farmer responded to the question of how decisions about what to plant and when were made with a laugh and said: "When Daddy says when." For some multifamily operations, every decision becomes a kind of power struggle between family members, as described by one farmer who had recently visited a neighbor's farm. "There's a neighboring farm which is a father and son operation. I was over there one day to get something and they had a cow in trouble calving and they had to have the vet come and pull the calf. They were trying to decide whether to keep the cow up on the lot or not. They were arguing about that. It didn't really matter if they kept the cow up in a pen or let it out into the field until the vet came. The son wanted to keep it up and the dad didn't. So they argued about what they was going to do with the cow!"

How families deal with short-term business decisions is a signpost of the underlying relationship between parents and children. It is an issue of the distribution of family power, especially how the father-son relationship has developed over the years. How power or influence is shared among the partners becomes especially problematic when multifamily farms confront long-term business decisions.

Long-term business decisions entail major investments of resources or fundamental changes in the nature of the business, such as whether to change commodities, whether to buy or lease additional land, or whether to adopt a new technology—decisions that are critical to the survival of the business and move beyond the questions of personal skills and knowledge into questions of goals and expectations. For many multifamily farm operations such decisions force members to confront the problem of children who need or want to "gear up" and parents who need or want to "gear down," as on Basin Spring Farm. Looking back on his partnership with his father, one farmer commented: "I would've made a few different decisions along the way if I had had full control of the farm. Of course that's the trouble when you are in a partnership with your father. He's wanting to wind down, to slow it down, but you see you need to be going forward instead of back."

Entering into a long-term decision is difficult, as any business operator will attest. Tension, risk, and uncertainty are attached to any major change in the business operation: Is this the right time? Can we afford to wait? Can we afford not to wait? When you add the potentially confounding factors of family relationships and partnership arrangements that are not clearly specified, the decisions become even more difficult. An issue that frequently emerges is differing estimates of risk and differing abilities to tolerate the uncertainties that come with major changes. "I've got a hog operation and it's very profitable. That is the only thing that I probably can ever remember doing against my daddy's will. He told me that I shouldn't do it. He said that it was too much money to spend to start up the operation and there was no guarantee that I'd get a return from it. But I just felt like that was what I had to do at the time, so I did it and it turned out all right."

For this farmer the decision to diversify into a hog operation in addition to the family's traditional grain operation was seen as a hedge against the future, a desire to ensure that one part of the business could provide a steady income. Individual goals and expectations often collide in business decisions, especially as they relate to what constitutes an acceptable income from farming. Parents sometimes complained that their child partners wanted to earn money too quickly and were not willing to wait, as they had. As a result, conflict often emerged over those long-term business decisions that entailed major investments of resources in order to achieve more stable or larger income returns. Studies have shown that money is the most frequently cited cause of divorce, and it is a similar source of dissension for multifamily farms. The struggle for control of financial resources and the financial decisions that affect the nature of the farm business, then, are frequently a point of conflict.

Older generations feel they have good reasons for being cautious about allowing younger partners to make significant decisions. The potential hazards of sharing financial decisions with children are acknowledged through stories of those multifamily farm operations brought to the brink of financial disaster. Unfortunately, such accounts are not apocryphal; rather they often are based in the experiences of neighbors and friends. One farm couple documented the rise and fall of their multifamily farm operation in this way: "Well, my son went to college, and then he came back to farm with us. He bought a bunch of land and a lot of machinery that he thought he needed. He kept saying 'We got to get big.' Farm prices were good in the beginning, but then they started going down. Then we had the drought and land prices dropped and he couldn't pay for everything. When we started, we had 230 acres; that would have been big enough for the two families. But everyone, like the government and the ag teachers in high school and college was stressing that you needed to get more land and bigger equipment and get over more acres with fewer hours with lesser hands. That is what got people like my son into trouble. He got us up to about 1,500 acres by the end. Now, we've sold off all but the original 230 acres to pay off the debts. Next week I'll be selling off the last of our dairy herd, and I'm not sure that will be enough."

Like many other farmers during the '70s and early '80s, this father had seen the family partnership go into bankruptcy and his son take a job in town, and he had been forced to sell off his dairy herd to settle debts. When we visited, they were facing the likelihood of having to sell the original 230-acre farmstead in order to finish satisfying the debts the multifamily partnership had accumulated. Encouraged by the predictions of experts, the upward trends in prices, and the successes of many others, the son had convinced the father to aggressively expand the family business and to "cash in" on the opportunities of the times. For these parents the desire to bring a child into the enterprise, to share decision making, and to acknowledge the value of their son's education led to financial disaster.

It is stories such as these that lead many parents to resist turning over full financial control of the multifamily operation to their children. Even though most father-son partnerships succeed, the tendency is to remember the failures. Moreover, stories such as these remind parents that if their child partners take financial risks because they are at the beginning of their careers and have the time to recover, failure may mean the loss of any hope the parents have to retire from farming. Such stories live on because they represent a justification for fathers to continue making the major financial and operational decisions.

But there are other reasons, ones that reflect a father's fear of losing his purpose in life and the difficulty of changing the nature of long-standing father-son relationships. As Barb noted, on Basin Spring Farm Jim was no longer "the boy" when he moved back, and this caused some disillusionment for both Jim and his father. Two other farmers speak to this difficulty of changing roles or acknowledging an adult-to-adult relationship, one based on equity between fathers and sons in farming skills and management know-how. The first farmer's comments are based on personal experience of how parents and children decide what to do with the farm and how they work together. The second quotation illustrates that it is not so much the suggested idea that may lead to tensions but who suggests it.

"I think my dad feels that turning the whole thing over would be just letting go of a whole lifetime. He can't reconcile to do that. He's not going to talk about it and he won't do it. I've told him he needs to makes some decisions about what's going to happen, but he says, 'I'm not going to make any decisions.' Yet when I do make a decision, he always has something to say about it."

"The father and son relationship on the farm can be a real trying experience, from what I've seen. My brother-in-law and me can talk over doing something new that actually is his idea. But if he goes to his dad with it, my father-in-law can't see it as a good idea. If I went to him with it he would agree just because I wasn't his son. Not that there is any difference between what I say or how my brother-in-law presents the idea, it just isn't a good idea if my brother-in-law suggests it, at least as far as my father-in-law is concerned."

There is great variety in approaches to business decision making on multifamily farms. Each business attempts to work out an approach that is most comfortable for the partners in view of (1) long-standing relationships and (2) assumptions about the alternatives and their outcomes that may be based on personal experience or observations of other farm families. However, it does seem that parents, particularly fathers, continue to exercise greater influence over business decisions, even when they have made efforts to include child partners. They may do so because of the difficulty that most parents encounter in acknowledging their children as competent adults. The results can be a strain in family relationships, especially when children attempt to assert their independence, and problems may arise later when the parent–decision maker is no longer there. This situation is illustrated by the following comments of a farmer who has been involved in a multifamily farm operation composed of the father and three sons in a loose partnership arrangement: "Like I say, you're looking at something that probably changed in the last couple of years. Up until Daddy passed away, there was no worry. Before that, the decisions were made by him. We always did what he said, Dad was in charge. Since Daddy died, I'd say it's not as smooth as it was. You've taken away the one that held everything together, and things haven't been as easy since then. It's a lot harder keeping things working so everybody feels like they are getting a fair shake."

The multifamily farm operation is a business, but it is also an intricate interweaving of family relationships. On the family farm, family members have a relationship to each other within the context of the family itself—spouse, child, parent—and a relationship to each other within the context of the farm business—manager, laborer. Sometimes the obligations and expectations associated with the family relationships clash with those of the work relationships. Then the service of family members as a labor force for the farm, which has been crucial to the economic survival of the American family farm, becomes a mixed blessing.

On Basin Spring Farm, Barb speaks of how well her husband and father-in-law worked together and the pleasures of having all the family pitch in to make the farm work. The same theme is expressed by many other multigeneration farm families. Others also note the special relationship that develops when parents and children and their spouses work together and recognize that the success of the business depends upon all their efforts. But it does not come easily. "It is like farming is a struggle enough without introducing the real problems. It is just the challenge of trying to get two generations together, to get one to recognize the rights of the other and to try to accommodate to each other that complicates everything else."

It is not easy to tell a family member that she or he is a poor worker, or that he or she has made a bad decision. For parents, it is not easy to balance an adult child's desire to provide a better living for her or his family with the needs of the farm business to invest in new equipment or land. For a woman, it is not easy to manage her obligations as housewife and as farm laborer, or as mother and as supervisor of children's work, or as daughter-in-law and as partner. For couples, it is not easy to shift from a marital relationship, where spouses see each other as equals, to a work relationship, where one serves as employer and the other as employee. For an adult son in multigenerational farm operations, it is not easy to tell his father that the work has become too hard or the machinery too difficult to use. And it is especially difficult to tell one's father that one is no longer a child, but a responsible adult.

Sorting out and defining these multiple work and family relationships can be especially stressful. It is not easy, and it may be impossible, for a husband to keep his frustration over his wife's inability to rake hay properly from coloring his feelings toward her when they go into the marital bedroom at night. It is not easy, and it may be impossible, to control your anger over your father-in-law's refusal to sell some cattle when your family needed some additional cash, when you all sit down for Sunday dinner. "Well the thing about a family operation is you kind of get on each other's nerves a little bit. You try to overlook it most of the time. If you weren't in a family business you would probably end up pulling each other's hair out or firing each other. You just have to learn how to deal with each other on a family farm, and this is probably one of the hardest things to do."

Particularly difficult may be the position of a daughter-in-law in multi-family farms. It may be difficult for her to define simultaneously the nature of her relationship with the family and with the farm business. Often daughters-in-law find themselves working side by side with their husbands, yet not having a voice in the decisions about the farm. Barb put it well, "My role as a daughter-in-law was very subordinate—to the farm and to the family, in that order." Although she acknowledges that, given her nonfarm background, this was probably appropriate, that does not make it any easier. Another daughter-in-law explains how she gradually carved a place for herself in the family farm: "In the beginning my husband's mother would work in the field and I would cook lunch and take care of the baby, and then she would come in during the afternoons and I'd go into the fields. I was just learning and I reckon they had enough confidence in me that they were willing to teach me. As my son got older, my mother-in-law began to let me do more and more of the farm work like I wanted to. I wanted to be out in the fields with my husband. I just slowly started taking on more of her responsibility and what she had done, and she took care of the boys. So it just started building from there. Now my mother-in-law she basically does the cooking and watches the kids and I do the farmwork."

Some parents are willing to allow a daughter-in-law to work their way into the operation as full partners. Others, for a variety of reasons, are not willing to allow this option, and nothing she does will be able to legitimize her in the role of a full contributing partner. For these wives of child partners, the farm as a business remains seen through a curtain. They know something is going on outside but they cannot clearly see what is happening, and they are not part of the action. One of the ironies of the position of many daughters-in-law on multifamily farms is that their sons come to be seen as more legitimate partners in the farm operation than they ever have been or ever will be. In such a situation the daughter-in-law must find a way to cope with her marginal relationship to the farm business.

Sometimes daughters-in-law accept the fact that they will never be a key player in the family operation because they may not come from a farm background or simply because they are not blood relatives. Sometimes they reject the struggle for a place in the family farm operation ("his domain") and build a place in the family based on an off-farm career ("her domain") and what this job contributes to the family's and the farm's economic resources. And sometimes they accept a marginal role of gofer or helper. In such cases the daughters-in-law may experience great stress because they are part of the family business, sharing in the responsibilities and the financial obligations of the business, but have little influence or authority over the decisions that affect its success. It often seems as if the only time the daughter-in-law is allowed to be involved is in a crisis, when she is expected to help solve a problem she had no part in creating. Moreover, there is always a sense that the work she can do or is permitted to do is not valued as highly as the work she is not allowed to do.

Besides working out the nature of family and work roles, multifamily farm businesses also must decide how to allocate moneys for family and farm needs. Farming is not a business that generates a lot of disposable income. Income does not come steadily in biweekly or monthly paychecks for most farm businesses, and all too often, when the market check arrives, it is not sufficient to pay outstanding bills *and* have enough left over to buy desired goods. This is especially true in the early years, when families are struggling to acquire the land, the equipment, and the livestock that will enable the farm business to survive and hopefully grow in the coming years. One farmer describes those early years: "It was kind of hard in a way when we were getting started because I started out on a quarter share of the crops, and of course I didn't have all that much money to put into buying equipment and the things we needed for the family. Every time we'd buy a new piece of equipment then I would buy in on that equipment until we finally ended up where we replaced everything. But it made it kind of hard then, because everything that we made and really needed to live on was going back into the farm to pay for our share of the equipment. It made it kind of rough there for a while."

On most family farms it is difficult to separate spending decisions for the home, or the family, from spending decisions for the farm, and it is even more complicated on multifamily operations. For some multifamily farms such decisions become a serious problem, because adult children have lifestyle expectations for themselves and their children that differ from those of their parents. What constitutes a necessity differs from one generation to another and from one family to another. Moreover, when partnerships are not formalized, it is not always clear who has the responsibility for providing or paying for the various items necessary for the farm business (for example, fuel, spare parts, service calls). Rarely is there enough money to satisfy everyone's desires or every family's needs, and so decisions must be made. Working out these spending decisions can be stressful because often the family partners do not agree on how much should be coming from the farm business to each family and how much each family should be contributing. "The first year we were in partnership with his dad, we noticed several areas of our expenditures that were just too high. We noticed little things like our fuel bill was unusually high. He got to paying attention to what he was doing. Too often he was bringing a piece of machinery over from his dad's place with an empty fuel tank and he was filling it up here, and then before he took it back he filled it up again. So we were providing more of the farm fuel for the family operation than his dad. That may sound nitpicky but it really made us aware of how we were spending our money."

Basin Spring Farm has been the home and the business of four generations of the Foote family. In each generation someone has wanted to farm, to continue the legacy. The Foote parents and thousands of others like them have worked to help their children attain their dreams. What they have been able to offer has varied, often depending on what they had to give. Yet even if it was only a word of encouragement, or the offer of loaned equipment, or family help at harvest time, it has usually been enough.

Multifamily farm operations require a delicate balancing act. Parents and children have to come to a new understanding of their relationships: parents have to be willing to allow their children to grow up, and children have to be willing to shape their aspirations in terms their parents can understand and appreciate. Family members must rely upon each other's labor, but they must also understand when that work is influenced by their other family roles. Family resources must be used to support not only the families and their households, but also the farm business. Family members must be willing to make trade-offs to insure that both the farm and the families survive. Such balancing adjustments are not easy, but they are necessary, as each generation knows.

5 The Economic Challenges of Farming as a Business

FOOTWORK

> Sometimes all you can do
> is light out
>
> stick in hand
> dog trotting ahead
> out the door, down the drive
> over the fields and
> into the horizon
>
> left foot, right foot
> whatever is propelling you,
> threatening to explode
> into blood and bone and raw
> emotion suddenly is
>
> at your back
> and all you feel is sweet
> air rushing over your cheeks
> past your ears and into your hair
> your walking stick a metronome:
> leftfootrightfootleftfootright
> sky gravel hill pond
> fescue cedar dog
> and You
>
> arriving
> before you get there.

BASIN SPRING FARM

When the farm crisis occurred in the early '80s, once again we heard people, especially government officials, talking about the unfortunate but necessary loss of farm operations that were simply marginal to the business of farming. We heard that the farm families who lost their farms to foreclosure were poor business managers who either had greedily expanded their operations beyond their management skills or never were really serious about the business of farming. They talked about farming as a business as if it were like any other business in America—take a big dose of good management, add a generous amount of hard work, a cup of smart investment, and a sprinkle of luck, and you have a successful farm capable of supporting a family. When others talk about farming as a business, there is an implied condemnation of those who struggle to make it, of those who believe that farming is more than a ledger book of assets, inputs, and debts. They talk about farming as a business as if it could exist independently from farming as a way of life, as if you could unravel where the one begins and the other ends.

This kind of thinking gives many of us fits as we wrestle with the terms of survival. When we came back to the farm it was to make a living at farming. Like Jim's family before us, we intended to make farming our business. We would grow our own, build it better ourselves, be self-sufficient from the bounty of the land. Jim's farm background made him more adept than I at taking new directions as necessity pointed the way. But my vision, what I wanted our life at Basin Spring to be, left me ill prepared to be sent plunging into the crosscurrents of survival. As a result Jim and I were often on opposite sides of the same fence: I getting up at six in the morning to garden before heading to work by seven thirty, and Jim wondering, "Why are you doing this to yourself?" Although I felt fortunate to have work that would contribute to our cash flow, I was dismayed and angry to know that my time was worth more off the farm than on the farm. What had happened to growing our own food for the table and raising pigs and chickens and rabbits and geese? What about the things that we had come home to the farm to do that we could not do anywhere else? How could we separate the farm as a business from the farm as our way of life?

I came to understand that the "crises" of the late '70s and early '80s were nothing new to either the life or the business of farming. Farming is about change, and farmers have always done what was necessary to make it work for them. We were simply experiencing another change involving what it means to make it in farming and what makes farming a business like no other.

Begin with Mother Nature. You simply cannot pencil an annual cash flow when dealing with the wind and the rain and the sun. The drought of '83 offers a graphic example; it gave me an introduction to the wrath of Mother Nature. It began with a long, wet spring that kept the farmers out of the fields until the second week in June. The last recorded rain fell on June 5, a sprinkle fell on July 3, and there were one or two other hints of what might have been rain after that. But for all practical purposes it did not rain for three months. Soaring temperatures accompanied by high winds turned the soggy, plowed ground of mid-June into hardpan littered with concrete clods by the first of July.

By mid-July we knew we were in trouble as the pastures turned brown. By August they were ashen, and the grass crackled under our feet like steel wool. Anything that managed to survive in the garden by August was stunted and gnarled; field corn was twisted from heat stress and fired yellow and brown at the bottom. Even the weeds died, leaving the corn rows and woods looking as though they had been mowed too close. Day after day the wind blew like the breath of hell. And when the wind finally lay, a quiet settled, as if the land was braced for death. It was, day after day after day.

It was called the worst drought in fifty years—since the Dust Bowl of the '30s—and I thought it would never end. I could not imagine what life would be like if it did. I would look out the back door and see our future burning up before my eyes. Would we recover from this? How? Surely the pastures would need to be plowed under and reseeded. And what would we do about the farm payments, clothes, shoes, health insurance, farm insurance, and . . . and . . . and . . . ? Finally I stopped looking out the back door and gazing out the windows. It was just too much. I turned inward.

That summer I began walking when I got in from work, to burn up the anxiety that roiled inside. I would hike to the back of the farm and along the creek that wasn't any more, as far as I could go, saving enough energy to get me back home again. As hot as it was, the fresh air helped, and the forward movement, and the dog and the cat trotting along for company. No matter how I felt coming down the drive in the afternoon past the seared tobacco, I looked forward to my walk before supper. I needed it.

About the time I had resigned myself to a life of blistering heat and dust, the drought broke. It was mid-September. By the first of October the pastures that I had given up for lost began to turn green. New tomatoes grew again in the garden, and the withered green-bean vines began to blossom. By November we were eating fresh spinach and kale. It was a miracle. And it was a lesson for me. A lesson that bloomed out of the searing months of the drought of '83. Life had been lurking at the fringes of hell, the miracle of life. It had been there all along, and it would always be there. But I could not know until I had lived to the other side of my first drought. And I had, we had.

We had a crop that fall, which made us fortunate by most standards. Our yield was thirty-three bushels per acre of corn and nine bushels per acre of soybeans. The tobacco, which had come within days of burning up completely, somehow made poundage. We had a crop, and I had survived my first drought. Another drought in 1988 was again called "the worst drought *in fifty years," but it was nothing like the drought of '83. Nothing could ever be, not for me.*

Mother Nature has offered us other extremes over the years. In 1976 a late frost on the tenth of May prevented grain formation, and we lost sixty acres of wheat, our entire crop. Sometimes a loss can be absorbed during the season, but that year we had to borrow money to see us through the summer. That was expensive green manure, turning under the wheat crop in the spring of 1976. But it was not wasted on the land. The spring of 1990 was wet, wet like never before. Six inches of rain fell on one afternoon in March. The bottomland along Sinking Creek flooded repeatedly, before and after tobacco setting, and by fall the crop was twenty-five hundred pounds short. It was our first shortage in sixteen years. That poundage could be added to the '91 allotment, like deferred savings. But there is no such provision for the other crops.

International politics is a fairly new factor in the business of farming and can throw a bizarre curve into the best-laid plans. In 1978 the Russians invaded Afghanistan, and the reverberations were felt at Basin Spring Farm when President Jimmy Carter slapped an embargo on Russian-purchased grain. The price of soybeans plummeted. The Russians turned to China for grain imports. We had to turn to the bank for the three thousand dollars that we lost in operating capital. Our net income that year was five thousand dollars. That was the fall that I took an off-farm job. In a similar twist of fate, Saddam Hussein decided to invade Kuwait in August of 1990. The price of gas and oil skyrocketed 20 percent, coinciding with the harvest season, but there was no subsequent hike in the price of yellow corn, which brought $2.27 a bushel that fall. Consider that yellow corn topped $3.00 per bushel in 1973 and brought $2.30 in 1977, and you get the picture.

The farm economy does not compensate for Saddam Hussein, or for skyrocketing production costs, or for the cost of living. It adjusts to what the market will bear. Period. The reason Americans pay less for food than any other country in the world is that farmers make it available, by whatever means necessary to provide it. Often it takes a piece of our hide.

Farm equipment is one of the most costly expenses on the farm. Our John Deere 1530 tractor with a loader cost $7,500 in February of 1976. We set it up on monthly payments for five years, and it was a great day when we retired that note. The 4,450-bushel grain bin cost $3,853 in May of 1976, and we put it on a four-year note. These are modest investments by most standards today, when a combine can run upwards of $100,000. But on any scale, farmers have to spend a lot of money to make a little money.

Your equipment has to be kept running, which is the flip side of owning farm machinery. The more expensive the equipment, the more costly the parts and labor. In the spring of 1990 both of our tractors were down at the same time and had to be loaded onto a flatbed and transported to the dealer. They were returned the following week with a nine-hundred-dollar bill. That same week, on Memorial Day afternoon, the heavy rains we had been experiencing for weeks caused a rock slide at the spring that severed an underground electric line. Our electrician advised that it would be cheaper to run a new line above ground than to try to dig up the problem. But that required sinking three new electric poles. The bill for that afternoon was one thousand dollars. An expensive holiday week at Basin Spring.

The expense of breakdowns multiplies when you are racing against time, especially in the spring and the fall. A rule of thumb for planting yellow corn in Kentucky is that for every day that corn is planted after the tenth of May, the cost to the farmer is one bushel per acre in yield. That may not sound like much of a loss, but if you are putting out several hundred acres in the first week of June, as happened in the spring of 1990, it mounts up quickly. You work under a similar strain in the fall. An early frost can shatter soybeans on the ground, so you have to keep your eye on the weather and be ready to run when conditions are right. You need to beat the November rains or risk prolonged wet spells between trips to the field. You want your downtime for machine maintenance to be in January and February, not November.

Then there's the issue of farm labor—to do it ourselves or to hire it, and what kind of arrangement to make if you do hire. We have done it every which way at Basin Spring. When we farmed full time we relied on seasonal help for jobs like hauling, housing, and stripping tobacco, and putting in hay. After we took off-farm jobs we added a full-time hand, but the convolutions of paperwork, filing social security and taxes became an additional burden. It was hard to find good help, it was hard to keep good help, and it was hard to keep up with them on paper if you did keep them working. We tried it for a couple of years and decided that we needed to simplify all around. We got rid of most of our equipment—the corn planter, the tobacco setter, and an International Harvester 205 combine—and let our farm help go. If we couldn't do it ourselves it just wouldn't get done—and it often didn't.

Then we were faced with the question of how to generate income without owning expensive equipment and hiring expensive and time-consuming labor. We talked with several farmers and found two who were doing an unusually good job. We considered a rental agreement, but we eventually worked into a fifty-fifty trade with one of them. He furnishes the equipment, the labor, and half of the out-of-pocket expenses to put out corn, soybeans, and wheat. We provide the land, a grain bin, and the other half of the crop expenses. Our tobacco crop is handled similarly, with a family who provides

labor while we provide the land and two tobacco barns. But because to-
bacco is a more labor-intensive crop, we furnish all the out-of-pocket expenses
with one exception: the tenant shares half the cost of the tobacco leases,
which we procure and handle through the season. The arrangement has
had complications, but basically it has worked well for several years to keep
costs down and to maximize production.

Change is the only certainty we have in making farm decisions: when to
sell, when to hold, when to expand, when to sit tight. The market fluctuates
with land values, a fact that crippled everyone's borrowing power in the '80s.
Property taxes rise, and in a good growing season—when everybody has a
big crop, as in 1990—you can count on the price of grain going down. The
cost of borrowing money is an on-going a challenge. Our farm plan changed
drastically in 1979 when soaring interest rates added three thousand dollars
to our farm payment right before spring planting and cost us our operating
capital. Low interest government loans help smooth some of these economic
wrinkles, as well as the natural (and unnatural) calamities that farmers rou-
tinely face. But a loan is a loan is a loan, and it must be paid back eventu-
ally—plus interest.

The family aspect of family farms offers one of the most complex vari-
ables in running a business, as attitudes and expectations are handed from
one generation to the next. The overriding fact of all farming operations,
however, is that the ways we have learned to survive have changed the way
we are living on our farms and in our rural communities. The trade-offs have
been plain to me over the last fifteen years. Sometimes I wonder if they have
been worth the price. But I have never doubted the way of life.

THE BUSINESS OF FARMING

What is a farm? This may seem like a silly question, since we all have an
image in our mind when we hear the word *farm*. But a farm means different things to different people. For government purposes a farm is an
enterprise that has gross sales of at least one thousand dollars in agricultural products a year. By the standards that we apply to most other businesses, this amount of sales represents a very modest economic activity.
But we are looking at a minimal definition; that is, at a minimum a land-
owner or land renter must sell at least one thousand dollars' worth of
natural commodities to be treated as a farm business. This minimal definition of the economic units in the farm industry includes an enormous
diversity in the size, the commodity mix, and a host of other characteristics that describe individual farms. For example, the government definition of a farm includes a twenty-acre operation that sells the appropriate

amount of pick-your-own strawberries in the spring and pumpkins in October, a three-thousand-acre wheat farm in Kansas, and an eighty-five-acre
corn-hog operation. The variation is most apparent in Kentucky, where
the geography and the physical environment permit a great diversity in
farm enterprises. Three of the farm couples we interviewed describe their
operations:

A dairy and tobacco farm: "We've got about sixty cows. We are down.
We were up to about ninety cows when we got into the dairy herd voluntary reduction program. We never have really built back up to what we
were since then. But we raise about thirty thousand pounds of tobacco.
We raise a lot of our calves, and then we buy some from other dairy farmers that we feed out and sell as beef cattle. On our three farms we have
about three hundred acres."

A grain and cattle operation: "We've probably got 500 acres of beans
planted and we've got about 350 to 375 more to plant double crop behind wheat if it rains. And we have something like 250 acres of corn out.
I have approximately sixty-something beef cows and about thirty-five year-
ling heifers that I'm keeping. And this spring I fed out fifty."

A subsistence farm: "We're not raising anything on our forty-three
acres but pasture and the sheep. We've got about sixty ewes. We don't
have any corn or anything like that. I've tried growing corn before and
had it flooded out in the river bottom. We also have a milk cow. In the
past years we've bought holstein bull calves, milked her, and then fed them
from that for about three months. Sometimes we'll have two or three
calves at a time. We are always turning them over, almost once a month
we are selling a calf when she is milking. The orchard hasn't produced
anything yet, but it will. All of the vegetables that we eat are put up from
the garden. Of course, all of the meat that we eat we raise. We also raise
chickens, and we sell eggs."

Is farming a business? It must be, for the farming families depend on
their ability to extract some financial returns from their labor, time, and
money invested in the operation. The fact that farmers produce a product that is sold on a commercial market makes farming a business. But it
is a business both similar to and different from most others in our economy.
The characteristics of farming as a business create an economic activity
that often is beyond the logic and rational dynamics of most other firms
in the marketplace.

From forty-three acres to over one thousand acres, from sheep to
cattle, from orchards to tobacco to soybeans to wheat to corn—such diversity makes it hard to talk about farming as a business because farming has
so many meanings, so many forms, and so many types of enterprises. We

cannot discuss the mythical "average" farm business, because it does not exist.

How similar is farming to other businesses? Some would argue that there are two key characteristics of the farm business that make it fundamentally different from other types of economic activities. First, the farm as a manufacturing firm tends to be place-bound, that is, you cannot pick up and move the farm somewhere else if the conditions of operation do not suit you. A related fact is that many commodities are geographically bound, that is, commercial grain operations are limited to specific climatic and landscape conditions; sugar cane does not grow in Minnesota, and blueberries do not grow in Florida. However, even these characteristics do not apply universally: it is possible to move some types of operations. For example, dairies are somewhat mobile, since what they need is a milk market and land for buildings and pastures. In the last few years commercial dairies have begun to move from central California to Idaho, and enclaves of former Dutch dairy operators are appearing in Texas.

Farms not only tend to be place-bound; they also are family-bound. As noted earlier, most people enter into farming through some type of arrangement with a family member, typically a parent, because of the enormous costs of starting in the business from scratch. As one of the farmers said: "If you had to go out and buy the land and this equipment you'd be talking about close to one million dollars. At one time it was way above that before the price of land came down." Therefore, persons who farm are likely to have been raised in farm families, and they are likely to be operating family farms or separate farms within the same general area as their parents' or other relatives' farms. These tendencies have become more pronounced as the cost of entering farming has risen. Most farmers recognize the family connection as the limiting condition on the mobility of farm businesses.

But is farming so different from other types of businesses in its lack of mobility? There are many family-owned and family-operated businesses throughout the American economy: retail stores, manufacturing firms, and professional and business services are all examples of economic activities that have been and continue to be important contributors to local economies. Such businesses may be transferred from one generation to another, and if a child were to decide to start up the same kind of business elsewhere, it, like farming, would require a considerable financial investment.

Tourist businesses that rely upon the attraction of a natural feature, like the Snake River, the Land between the Lakes, or the Florida Keys, are also tied to a particular place. You cannot pack up and move your charter boat operation when the fishing is bad or move your rental canoe operation when the river is low. Other businesses, such as professional services, may be established anywhere, but become place-bound by the trust clients develop in your professional skills and the reputation you have built in the area. It is possible to move and begin again, but it involves an extended period of rebuilding the reputation and trust you left behind. So there are other types of businesses that, like farming, are family owned and tied to the specific geographic features of an area or can become place-bound over time.

Managing a farm as a family-owned, family-provided business is more difficult because of the total dependence on weather as the critical production factor. It is often said that everyone complains about the weather but no one does anything about it, but there are few other economic activities that rest so completely on the whims of nature. Tourism and construction both rely upon good weather, but in different ways and to a lesser degree. Heavy rains slow construction, but they do not have the potential to destroy the business income for the year. If the snows do not come, the ski season and the tourist dollars may disappear. But most resorts now have the ability to make snow, which may not be as desirable as new fallen snow but will still bring the skiers to the resort. Droughts can be devastating to water-based recreational businesses, but again, the effects are episodic and it would require a very long drought to seriously affect annual income. Thus, although the weather may be a limiting factor for many other small businesses, most have the ability to work around these limitations and modify the effects of the weather. Farming is more quickly affected by the weather, and the strategies to reduce the effects of bad weather are more limited.

The dependence on something out of your control for the very survival of your business, combined with the difficulty of simply moving and setting up somewhere else, defines the essential insecurity and the stressful nature of the farm business. During the drought of '88 we would visit farms where the expected rotational planting of soybeans, following the harvest of the winter wheat, had begun. Clouds of dust darkened the horizon as farmers, with faith in the possibility of rain, planted their soybeans. One of these said: "I was planting beans today. When I'd get to one end of the field I'd have to stop until the dust traveled on so I could see where to go back. It was so dusty you couldn't see. I've always believed it would rain. It always has. But this time it may not rain on time. But even feeling this, if I don't plant the beans I know I won't do any good. I probably wouldn't plant them if I hadn't contracted some back. If I don't plant them I can't fill the contract. All you are running on is just hope, that's all. It's not a pleasant feeling to be out in the dust that way planting beans."

In order for a farm to succeed, the rains must come at the right time, in the right amount, and in the right way. The weather is critical for the growing of crops, but it also is a factor in the development of the viruses and blights and pests that can destroy a crop or a pasture or sicken your livestock, and lead to substantial economic losses just as easily as floods or drought or frosts. If there is too much rain and humidity, even though you may have a good crop of tobacco, blue mold will develop and you lose the crop anyway. There is a host of weather-related blights and illnesses that spell the difference between a good year and a bad one, and the fact is, the farmer has little or no control over any of them. The frustration of watching your financial plans dry up or wash away is evident in the following: "This year, up until it started raining, we didn't have any pasture at all. It looked almost like a desert, it hadn't rained for so long. Last year was almost as dry. This year we got one inch in May and no more until it started raining in July. I don't raise much corn, just enough to feed my cattle, but it was a total loss; it had dried up. I cut my hay, but all I got from the first cutting was about half a crop. As soon as I cut hay I had to start feeding the cattle because the pasture was completely gone. I got scared and was afraid that it wouldn't rain, so I bought some of that Oregon hay they were offering around here. I figured my tobacco would just about pay for the hay. Then it finally rained. We had close to fifteen inches in a little over a week. It would come five to six inches at a time. Monday here we had plenty of rain. It rained half an inch down the road and I guess it rained close to five inches or more in our area. We just happened to be in that strip where it decided to rain. There was wind and hail too. It rained so much that about half of my tobacco crop scalded down from too much water. I won't even make expenses from what is left of my tobacco. The only thing I had to pay for the hay with was my tobacco crop, and it rained so much that I've lost most of it."

So the rains came—too late for the corn, on time for the pastures, and too much for the tobacco. What began as a year that on paper figured to be profitable ended with more debt to pay for the hay that fed the cattle, next year's potential profits, through the winter. And so it goes. If weather patterns shift and your area no longer gets as much rain as it used to, you cannot close the farm business and move somewhere else. The farm does not move. So you make adjustments; you borrow the money to set up an irrigation system, if you are lucky enough to have a nearby water supply. You continue farming, but at a higher cost of production than before.

One of the ironic mixed blessings of this situation is that drought in the Midwest is a boon to the corn farmers of the southeast, and vice versa. Southeastern corn farmers profit from the drought-induced higher prices for corn, while their fellow farm business operators in the Midwest endure the costs of a lost crop. When cattle prices drop, cattle farmers do make smaller returns, but hog farmers pay a price too, as their product becomes less competitive in the supermarket. It is the nature of the agricultural marketplace: good years of high production translate into lower prices for the producer. Droughts and floods mean losses for some but greater profits for others.

Production in farming is sequential rather than continuous. What this statement means is that in farming there are extended periods of time when a minimal amount of labor is required, and then there are compressed periods of time when significant amounts of labor are needed. In most businesses labor is required on a continuous basis, so employers can assure year-round employment to their workers. Or, in some seasons, like Christmas, you may need more workers temporarily, but your profits for the entire year will not be lost if you cannot get all the workers you need. The sequential nature of farm work makes managing the supply of labor difficult.

In the past, when the alternatives to farm work were few and far away in urban areas, farmers could rely upon a pool of laborers who were willing to work for limited wages and in-kind payments (for example, housing or a cow or a hog) and were available for seasonal work. But today these laborers have gone to the cities or work in rural factories. In the past high school students would work during their summer vacations putting up hay or working in the tobacco. But today there are fewer teenagers in rural areas and the local fast-food franchise pays more for fewer hours in better conditions. In other cases the welfare system operates so as to penalize recipients with loss of support if they work too many hours for pay. Since so few farmers can offer fringe benefits, such as health insurance, to compensate for their low wages, the balance sheet for welfare versus farm work is weighted away from farm labor.

As a business, farming is facing a growing labor crisis in terms of the availability, the quality, and the cost of hired labor. The crisis begins with the competition with factories and other industrial employers for qualified workers and the simultaneous increase in the technical complexity of farming. Factories, in general, offer workers better working conditions, continuous employment, and better pay and benefits. The technological revolution in farming means that workers need to be literate and to have a higher level of technical skills than in the past. A farmer describes this competition with other local employers for hired labor: "Good labor is hard to find. Your good labor is already in the factory working eight hours a day and getting paid ten dollars an hour or more. But here, all I can pay a guy is five dollars an hour, and he will work twelve to fifteen hours on a busy day, and for that he gets no benefits. Most people just like an eight-hour day, and I don't blame anybody for that. It's just harder to find anybody that wants to do farm work and can handle the equipment and are dependable. I would say that labor is the number one problem and the number one expense. In order to get a man that won't tear up a ten-thousand-dollar piece of equipment, he's got to have some sense. But you've got to pay him the price. Look what you're competing with: benefits, like health insurance, they've got two weeks paid vacation and they're starting out at ten dollars an hour with nothing but maybe an eighth- or tenth-grade education. As farm people we just can't compete with that."

So, even when you can find qualified workers who are interested and willing to work on the farm, in order to attract and keep them, you have to pay a salary that is at least near what they could earn at a town job. All employers face the problem of higher labor costs. But farmers cannot figure these higher costs into the price of the products they sell so as to maintain their profit margins. One farmer sums up the situation this way: "Everybody that is in business in the U.S. says that they want so much for their product. They put out a product and say this is what it costs me to produce it—I have to have so much for my product. But when we take what we produce to the market we say, 'What are you going to give me?' We can't get more for our tobacco because our labor costs went up or more for our hogs because the cost of our feed went up."

The competition for labor with factories and urban employers is compounded by other problems: the combined effects of the social service system, which reduces support payments dollar for dollar for earned wages, and the income tax system, which demands concrete evidence of the payment of wages. It is possible to find people willing to do seasonal work, but often they only want to be paid in cash, and tax laws no longer permit a cash-basis business. "We used to have a list of seven or eight names; anytime we could call them. The last year to a year and a half, especially your casual labor, has really dried up. If you will pay them cash you can get all of the help that you want. They draw welfare and food stamps, and they don't want that two- or three-week job to make them lose their benefits. Of course we don't want to pay them in cash because the IRS only lets you show so much in labor costs in cash before they won't let you deduct it as an expense. But we do pay some in cash because that is the only way that we can get them. You have to sometimes. You don't have a choice."

When you can find someone willing to do farm work, the financial arrangements open to farmers are more varied than those available to other businesses. Farmers have always been cash poor, and throughout history farmers have paid hired hands with a mix of cash and in-kind benefits. On the frontier of the Great Plains, these in-kind payments included housing, food, and often clothing. Today, farmers continue to devise new ways to find and keep the workers that they need to keep their business running.

Providing housing and other noncash benefits: "The couple I have working for me now were living off food stamps and welfare in an old house they couldn't keep warm. One day she asked how much we'd give her and her husband to come and work for us. I said I'd start them out on sixty-five dollars a week and we'd buy a trailer for them. We also paid their electric and water bills, and we give them two hogs a year. They have been here three years, and now he gets seventy-five dollars a week. His wife helps in the tobacco and I give her the same as I do her husband when she works. We give them two hundred dollars extra when we finish cutting tobacco. He very seldom works on Saturday and never works on Sunday, but I sure do."

Sharecropping with tenants: "Really, a tenant as far as sharing in a percentage in the crop really comes out better than the owner because they don't have anything but their time invested. They don't have the costs of upkeep on machinery and buildings, or the costs of insurance and taxes. I bought a new trailer for my tenant two years ago. I provide a place to live and a basic salary, plus he is paid an hourly wage when he milks and when he makes hay. I also give him the land for his own tobacco. This year he will raise five thousand pounds of tobacco on his own. My tenant doesn't furnish anything but himself. He is on vacation this week in Florida. Can you imagine a tenant having a vacation in Florida? When I was a tenant we didn't take vacations. But that is how times have changed. I stay home and my tenant takes a vacation! But a good tenant is hard to find and the one I have now has been here for four years. It is rather expensive to keep a good tenant, but I couldn't operate this farm without him."

Sharecropping and tenants have formal, historical meanings, but farm owners use these terms interchangeably. What is important to the farm owner is to find a way to attract and keep good labor and to do so in a way so that their costs do not exceed the economic benefits from that hired person's work. It is a delicate balance, and sometimes there is a sense that the hired labor gains more from the relationship than does the employer. These farmers' stories illustrate the mixed blessing of finding and keeping good labor; one often wonders whether the costs are too high, and who is benefiting most from the arrangement.

Even when you find someone to hire and you arrive at a mutually satisfying financial arrangement, there remain problems, as any employer knows. Sometimes those who are willing to do hired farm work are willing because they do not have the skills to compete in the industrial market. Matching workers' skills to the jobs they need to do is not easy. The nature of farming today—highly mechanized, with a heavy reliance upon chemicals to extract the maximum amount of productivity and upon the careful balancing of feed to produce the most weight gain or milk production in livestock—demands technical skills on a par with those required of employees in other types of industrial jobs. Farm work is technical work, and it is not enough simply to find workers; you must be able to find qualified workers. Either way the costs can be great. This farmer describes the costs of finding good labor, and what happens if you do not: "We have help now that makes more money than we do, dollar for dollar, that is. But we went on and hired him because, if you don't have good help, you can ruin dairy cattle in a month's time. Somebody can come in and ruin a herd of cattle that you have been breeding for thirty years. One time I went to the hospital for some tests. I had someone else working for me then. I was only gone three days, but by the time I got back I had lost three cows. So that was a real expensive trip to the hospital for me. That is what you get into if you can't find a decent hand. Now my neighbor, he's got full-time help that have no education whatsoever. Yet he'll put them behind machines that are worth a lot of money. If they tear it up, and they do get tore up, it may cost him a thousand dollars or more to make the repairs. He's got a big operation, twelve hundred acres of grain and he milks one hundred cows. He can't do it all. But at the same time he can't afford to go out and find good help that would take care of things. I know the cows don't get fed right, he knows they don't get fed right, and he knows his business suffers from this."

Finally, as other employers recognize, employees do not have a proprietary interest in the business and so may not be as careful or as attentive to their work as the owner. So the work may not get done as you would do it; you may spend as much time supervising your hired help as they do working; and sometimes the work simply does not get done at all. "A lot of times the hired help don't pay attention to things that they should. As good as this boy is that I've got that works for me, I have to set up the spraying equipment for him even though we have set it the same way year after year. And he's way above average as far as help is concerned. But yet, when it comes to some little things that he really should know, he don't. I have to do a lot of the technical things for him, but once I do them he goes right on with the job. Don't get me wrong, he does a good job, he's careful, and we're tickled to death with him. But there are a lot of little things that he should know or do and he don't. He just doesn't care like I do. That's what you run into with any farm hand."

Although machines have reduced some of the harshness of farm work and reduced the reliance on hand labor, there are some jobs that machines do not do and some commodities, tobacco for example, that still require enormous amounts of hand labor. The continued importance of physical labor in all weather conditions adds to the difficulty of finding good labor. Central Kentucky is known worldwide for its burley tobacco, and the following description of the cycle of burley production illustrates that hard hand labor is still important on a burley tobacco farm.

Tobacco seeds are planted in March or early April in beds that are covered with canvas to protect the seeds. As the seedlings grow, the beds must be thinned by hand to allow the strongest plants to survive. In May and early June you move through the beds hand-pulling the plants and placing them into bundles, which are taken to the tilled fields. Tobacco setting requires at the very minimum three people, one to drive the tractor and two persons to ride the setter. But since most farmers have double row setters, you usually need at a minimum five persons to set, in addition to those who are pulling plants and making bundles. Tobacco setters take the hand-tied bundles of plants and place them into the setter, which plants and waters each one. In the years before chemicals you would have to move through the fields pulling off tobacco worms by hand. But today the next major work is done in early August, when the tobacco is topped, that is, the blooms are removed to force growth into the leaves. As you top, it is also necessary to remove any suckers (long, spindly growths) that have emerged on the early flowering plants. Today chemicals have replaced the time and labor once committed to suckering tobacco, but good farmers still go through their fields to check.

In late August or early September, twenty-one days after topping, the tobacco is harvested. Harvest requires the greatest amount of labor. The tobacco plant is cut by hand with a machete, and the plants are speared onto stakes and left in rows in the fields for a short period of time. The staked plants are then put onto a wagon and taken to a barn for "housing"; that is, they are hung from rafters to air cure. Cutting must be done within a short period of time so that all the plants are hung at the same time and therefore have the same time and conditions for curing. For this reason harvest requires many hands to get the tobacco cut and out of the fields. In late October or early November the hanging tobacco is brought down and the leaves are stripped from the stalks and hand tied into bundles. The bundles are placed into baskets to take to the market, which opens every year on the Monday of Thanksgiving week. A tobacco baler and press is now available that eliminates the hand tying of bundles, but some farmers remain concerned that bundled tobacco does not bring as much at market as hand-tied tobacco does.

What is apparent from this description of burley tobacco production is that there are periods of time when many people are needed to work in the fields or barns and periods when the labor need is considerably smaller. With most crops, as for tobacco, timing is everything. In order to bring the maximum price at sales, corn must be harvested when it has the proper moisture content, and the same is true of commercial hays. Other vegetables are even more time dependent. If you wait too long, your cucumbers are too big and bring less than half the price of smaller ones; and a day lost because you cannot find any help can translate into substantial differences in market price.

Like all employers, farm operators who must rely on hired labor feel that this issue is one of the major challenges in their business. The very nature of farm work—hard physical labor that often entails ten-hour or twelve-hour days to beat the changing weather, often in hot and humid conditions—means that few potential employees find it attractive. Obtaining labor that is qualified to operate heavy machinery and also capable of handling livestock and mixing feeds and chemicals in proper ratios at a cost you can afford is a continuous struggle. The owner of a dairy in Florida that was milking twenty-four hundred head three times a day and therefore running three eight-hour shifts of workers to do the milking, manage the herds, and maintain records, offered the following observation. When asked what was the factor that would most influence whether he expanded his operation, he said simply, "Labor." He went on to say that supervising his ninety employees, maintaining morale, and solving labor problems were the single most difficult part of his job. Most other business operators would say, "Amen."

For farm businesses labor is not easy to find, expensive when you find it, and not always well qualified for the job to be done. The same is often true for other businesses. But many farmers have the added problem that they need labor only at certain times and cannot afford to keep an employee or two year-round, when they really only need that labor for four to six weeks out of each year. Most farmers have no training in personnel management and employee relations, yet they find they must learn

to function in these roles. And the situation is becoming more complicated by new federal and state regulations regarding farm labor. Farmers
now find themselves in the same position as most other small-business
owners, managing their labor according to government regulations. It is a
problem of balance—a delicate balance between needs and resources—and
striking the balance depends upon several other factors.

An enduring truth about the business of farming is that farmers buy their
inputs on a retail market and sell their products on a wholesale market.
This practice, of course, is just the reverse of how other production operations do business. What it means is that farmers are price takers when
they purchase their inputs *and* when they sell their products. "The problem with farming is you take what they offer you and you give what they
ask. Whatever you buy you give what they charge. Whatever you've got to
sell, you take what they offer you." Therefore, in farming there are limited
opportunities to maintain a steady income from year to year by adjusting
the prices you receive for your products to compensate for changes in the
prices you pay for the inputs and labor that permit you to produce the
products. As a result, over the years farmers have watched the prices for
their inputs rise steadily, while the prices they receive for their products
have remained the same and in too many years actually declined.

Throughout this century farmers have come to rely more and more
upon purchased inputs. You do not save back seed corn for next year; you
buy it. You may raise the feed for your livestock, but you also purchase
feed additives to maximize the weight gain in your livestock. Simply put,
farmers are consumers as well as producers, and in 1986 they spent $122.2
billion for purchased inputs, a whopping 251 percent more than they spent
to operate their farms in 1964. This increasing reliance on purchased inputs, especially machinery, seed, and a host of new chemicals, has shifted
the equation of costs for farmers. Once labor represented the biggest expense of farming, but today, as shown in figure 1, capital represents nearly
seventy cents of every dollar spent for production. Capital is money, the
out-of-pocket moneys that must be spent for all sorts of manufactured inputs that are now essential to the operation of the modern farm. And as
the reliance on purchased inputs from fewer and fewer suppliers has increased, the prices have also risen. "I know when we milked about thirty
cows and sold Grade A milk, we were getting about $5 to $6 a hundred
pounds. It was easier to pay bills with the $6 milk than when it got to $12
and $13 a hundred pounds. One reason was that we did most of the work
ourselves. But now with all the mechanization, we have to pay the cost of
operating those machines. The same electricity we used to get for $65 a

Figure 1. Change in Input Shares to U.S. Agriculture: 1910-1986

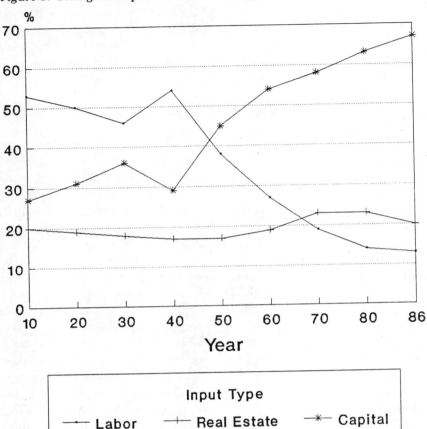

Source: Adapted from Tweeten, 1989, Table 1.4.

month got up to $125 or more a month. We raise a lot of feed for our beef cattle now. Diesel fuel used to cost about $.19, but a few years ago it got up to $1.00. Fertilizer has got so high. And the machines. To run over five hundred acres of grain you need a machine that costs around $50,000. There is no way to pay for all this. We have to pay $24 an hour to get somebody out here to work on our combines when they are down, and a farmer can't pay that kind of wages because he doesn't make that kind of wages."

While the prices of inputs have spiraled steadily upward, pushing up the expenses of farming, the prices that farmers receive for their commodities, their income, seem to have remained stable. The gap between the prices they receive for their products and the prices they pay for those things that are essential to produce their goods creates a sense of overwhelming frustration. There is also a sense that suppliers set their prices based on the peak prices of commodities, whereas declines, if they come at all, lag far behind drops in farm commodity prices. A farmer explains the operation of the marketplace from his perspective: "Several years ago the price of beans got way up there. But then all of our prices went up, everything we had to buy—chemicals, fertilizer, and equipment—is going to go up. Even if we don't sell any ten-dollar beans, if we end up selling six-dollar beans, all the supply companies are going to base their prices on ten-dollar beans. So we're going to get charged higher prices for what we buy whether we get the ten-dollars a bushel for the beans or not. I thought maybe twenty-five or thirty years ago that the farmer would get what was coming to him. But I don't believe that he'll ever do it. I don't believe that you ever will. Take all of your small grains, your corn, your cattle, with the exception of tobacco, there's not much difference in the price now than it was back in '45 or '46. Take oil—oil was cheap and gasoline was about five gallons for a dollar. Oil got to about forty dollars a barrel when it was high. Now it is fifteen or sixteen dollars, but gas hasn't gone down that far. And think about the prices of everything else. We got high prices for fertilizer, feed, taxes, water, lights, just everything. Everything that we have to have has kept going up. The things we had to sell kept going down."

An economist will tell you that the demand for food is inelastic, meaning that prices for farm products are relatively resistant to typical market forces. A good year with a bumper crop often means a farmer simply has more to sell at a lower price, with the net result that the farm produces about the same amount of income as it would in a year with normal production levels. In a bad year, when the harvest is small, prices may rise, but then the farmer has little to sell and so has limited opportunities to "cash in" on the higher prices. Here is a farm wife's explanation of this catch-22: "You can never get it right. You just can't. If the crop is good, then everybody's crop is good. If the stuff is so cheap you can't get rid of it, it's a bad year. So, you feed it through an animal instead of selling it at the mill, and you might do fine. We did that one year. We raised the best corn crop one year. People from the university came and looked at the crop. It was one of those one-hundred-bushels-an-acre crops. They were great big ears. It was beautiful. Corn didn't bring anything that year, naturally. Everybody had a great corn crop. So my husband decided we would buy some pigs. We would feed these little guys the corn. In exactly four months they all weighed 240 pounds. Pigs were up and we sold them and made good money. Then we turned around and bought little pigs again, because pigs were up, right? We thought that all we had to do was do this again. But by the time they were 240 pounds, pigs were down. It didn't work out at all. So then we went to raise another crop of corn and the blight hit it and so on and so forth."

What all this means, at least from farmers' perspective, is that they must spend a lot more to produce a lot more, but their net income seems to change little. Figure 2 illustrates that this frustration is real; it is not just a figment of their imagination. Without question, gross farm income jumped during the '70s, but so did farm expenses, and the net effect of all this was that net farm income remained the same or actually declined during the '70s and '80s. And although the gap between prices paid and prices received has always been there, it seems as if it has been widening over the years. Farmers are frustrated by the fact that their operating costs keep rising but the prices they receive seem only to change for the worse. Even when they try to keep their production costs down, it makes no difference in their income at the end, because they have no control over the prices they receive. The result is an anger and frustration shared by most farmers.

For farmers, the decision to absorb these rising costs of operation is no choice at all. For example, as federal and state regulations on milk production have increased to protect public health, dairy farmers have had to mechanize their operations, increasing their use of electricity to meet regulatory standards. Purchasing commercial feeds for their dairy herds is often not a viable option, given the cost of such feeds, so most dairy farmers raise their own feed and bear the costs of the inputs necessary to raise it. Ironically, given the price of raising or buying feed, dairy farmers find it costs more to feed their own heifers to an age at which they can begin to produce milk than it does to sell them after they are born and buy them back in two years for dairy production.

The final irony of the cost-price squeeze, at least for farmers, is that they feel consumers blame them when the retail cost of food rises. News coverage of the drought of '83 always included a statement that supermarket prices would rise, and the coverage of the Midwest floods in '93 and the simultaneous drought in the southeast stated the same thing. They were right; supermarket prices of food did rise. When the price of beef rises at the stockyard, the media tells us the cost of hamburger is going to go up; and it does. But farmers feel that their share of these increasing supermarket prices does not change. Figure 3 indicates that at least through the '80s the farmers' share of retail food prices and total food expenditures changed little. Yet when the prices they receive decline, there is little

Figure 2. Farm Income and Expenses: 1970-1990

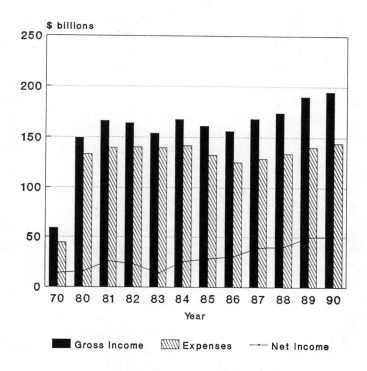

Source: Statistical Abstract of the United States, 1992: Table 1093

Figure 3. Farm Value Share of Retail Prices and Expenditures: 1980-1990

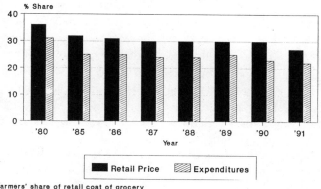

Farmers' share of retail cost of grocery
store food basket and of expenditures
for all foods purchased

Source: Statistical Abstract of the United States, 1992: Table 1101

drop in the price of food products at the grocery store. As one farmer explains, "You take your hog prices, back when hogs got up for the first time at about sixty-three cents, bacon went to one dollar a pound. Since then, hogs has been about thirty to thirty-five cents, but your bacon has never gone down." The irony here, of course, is that the subsistence farm family raising all their own food is now a myth. Farmers buy most of their foods at the supermarket just as everyone else does. So the gap between what they receive for their farm products and what they pay at the grocery store simply intensifies the economic pressures on their businesses and on their families. They are, in fact, paying more to earn less and then paying more to keep their families in food.

The cost-price squeeze is real, and the screw is tightened by the natural rhythms of production that determine when farmers will have products to market and thus what will be the cash flow in farm businesses. Unlike other businesses, where there is continuity of production and therefore a continuous flow of products to be sold, farmers are tied to the natural cycles of growth. As a result there are extended periods of time when many farmers have nothing to market. Or when they do have something to sell, the market is not always in their favor. As one woman farmer explained: "I raised beef cattle; I was cross-breeding Angus and Charolais, and they were wonderful. But it takes forever. It takes a cow a couple of years to be old enough to even date! Then it takes them nine months to calve. Then you have at least six months before you can do anything with them. I've heard this said: 'Oh, they can always sell something.' Sure we can. But it is never going to be ready. There is always a bill due before those cattle are ready to sell. So, if you have to sell out early, then you are not going to get a full return on your investment. You are going to make some money but not as much as you could, so that's a loss. You're forced to sell before the right time because your cash flow doesn't match your flow of debt."

Another farmer offers an insightful comment on the relationship between costs, prices received, and the timing of production. His comment illustrates how these factors influence cash flow on a typical farm. "I was over at a car dealer to have something done on my car. When I got the bill it was pretty high and I said 'Shew!' He said, 'You take veal calves and bring seventy-five to eighty dollars a calf.' I said, 'Yeah, seventy-five to eighty dollars a calf. How many calves do I sell a year?' He didn't give me an answer. A lot of people think that farming is a great thing. You want something, you sell a cow for six hundred to seven hundred dollars or a heifer for eight hundred or a thousand dollars. But it takes you two years to grow that heifer. You don't sell them every week. Plus you got to think about what it costs you to raise that calf to where you can sell it. People who aren't farmers don't think about everything you have to pay for out-of-pocket first to raise that calf before you get it back. You have your fertilizer, seed corn, and stuff like that that's high. You just don't have enough money coming in regularly to keep up with your farm expenses."

Outside of farming, they call it venture capital, the capital you invest in a business enterprise to start it up, to make the product you will eventually sell to turn your profit. Independent housing contractors often build a home without having a buyer in hand; movie producers spend millions to make a movie they hope the public will come to see; restaurant owners must make food in anticipation of the evening dinner crowd; and research and development companies invest millions to develop new drugs or biotechnology products with no certainty of success. In this sense farmers are like other venture capital businesses, investing substantial sums of money to bring a product to market. But at another level the typical farmer does not have a group of investors who have backed his or her venture. The farmer experiences the risks and the uncertainty alone, as an individual, whereas most other venture capital businesses experience risks at a corporate level.

With a venture capital enterprise, what you have to sell takes a long time to grow or develop. During this time you are making investments in its development. The result is a business enterprise that, by and large, has long periods of no income but bills that must be paid. For grain and tobacco farmers, payday comes once a year, in the fall or early winter when they sell their crop. Although grain farmers can hold back a portion of their crops in the hope that prices will rise, doing so entails storage costs and assumes that they have enough of a financial cushion to be able to survive without the income from their entire crop. Cattle and hog farmers may sell at any time of the year, if their livestock is ready. On the other hand, dairy farmers have a regular income. Milk checks come every two weeks. This situation is what leads to the often-heard comment that farmers are assets rich and cash poor. "You've got bills like everybody else. But when you are farming, you've got more bills than most people do, and less cash flow. We are not certain about where our next check is coming from, whereas people with public jobs know at the end of the week that they are going to get that check. We've got payments to make on machinery. We haven't got the land paid for. In the summertime, you've got to borrow money to buy feed, buy your fertilizer and the other chemicals you have to have to put your crops out. You've got fuel bills and utility bills. If your truck motor blows up you've got to put a motor in it so

you can get your cattle or your crops to market. Sure we've got land, equipment, and livestock, but that doesn't mean we got money on hand. You
can't always change those things into cash when you need it."

The cash-flow problem is serious. More bills arrive than ever before,
and everybody wants to be paid now. In many areas the days of buying
"on account" have disappeared. The depression of the early '80s in farm
country sent too many farmers with outstanding debts with their local
suppliers into bankruptcy. As a result farm supply companies found themselves struggling desperately to stay afloat, and some went under with
their farm customers. For many farmers, conditions of doing business have
drastically changed as a result. "The farmer is faced with more financial
stress than he ever has. On top of that, it used to be a standard practice
that you bought your seeds and fertilizer and whatever, and you paid in
November or whenever you sold your crop. Any more, you have to pay
that in thirty days. They don't want to wait and you can't blame them.
Farmers have gone broke and then the feed mill guy doesn't get paid.
They need their money just like anybody else. But this means that now
farmers have to come up with money that they don't have. This means
by the time money comes in from the crop it has already been used, it
has to go to the bank plus interest."

To a certain extent, as the technological complexity and the level of
mechanization have increased, as the costs of production have risen, and
as profit margins have narrowed, the zone of survival has shrunk. The
myth that only those who are poor managers go bankrupt and lose their
farms is not true today. Rather, current farm crises have been more demo-
cratic in their effects.

Social scientists now talk about the "bi-modal" distribution in farming due to the "disappearing middle." In Kentucky and many other states
the numbers of very large and small farms have been increasing, while
the middle-size farms have been declining. These family-size farms have
been either getting bigger or getting smaller or simply going out of business. There are many farmers who have done quite well. They have been
lucky enough to make the right decisions at the right time, such as expanding their operations when interest rates and land prices were low, or
they had little debt when the most recent farm crises began in the mid-
'70s. Other farmers, like Barb and Jim, saw the writing on the wall and
gave up the dream of supporting the family by farming alone; they cut
back their operation and took off-farm jobs. But many have not done well.
Sometimes their options have been limited—they had high debt loads,
there were no job opportunities nearby, or there was no land available for

expansion. But all farmers face business problems that will not be solved simply by hoping that the costs of inputs will go down or that the prices they receive will go up, and so the search for other solutions begins. The next chapter continues the story of the business of farming by examining some of the solutions that farmers have tried in order to improve their economic position.

6 Farming as a Business: Strategies for Success

BASIN SPRING FARM

One thing about farming that has changed in Jim's lifetime is the government. Government farm programs are a big part of our farm business today, but they are so complicated to keep up with, that it seems we are farming paperwork instead of the land. Unfortunately, despite the number of programs and their complexity, they only begin to compensate for the vagaries of the market and the weather. Sometimes they are a trade-off. The Payment-in-Kind (PIK) program paid farmers to take cropland out of production in 1983 and ended up rescuing many from the ravages of the drought. That was a good trade. But the ten-year set-aside program may not be such a good deal. It is a hedge against the rising costs of production, but if land values continue to rise, and if grain becomes more valuable on the world market, it will not be a good trade to have productive land lying fallow. The problem is trying to predict these values.

Government intervention extends to the farmer yet another double-edged sword. Although federal controls protect the market by monitoring production and providing a safety net of support prices, they require adherence to particular production standards, which in turn require meticulous records. The regulations are ever changing, not just from season to season, but within seasons. We have used these programs, and sometimes they have been the difference between breaking even and losing money. But we must be aware of every new wrinkle in the regulations, keep an eye on every deadline, and keep the paperwork flowing through proper channels, just to be able to participate, and it requires a vigilant eye and attention to detail.

Many of the benefits of the alphabet soup of programs are clear cut. Programs such as soil conservation incentives, low interest loans, and disaster insurance seem initially like good ideas. In 1982 the PIK program provided the lucky ones who participated with money for not planting a crop that would have been wiped out in what was regarded then as the worst drought in fifty years. The downside of government programs is control and submission. Many farmers dislike having the government control what and how much they will grow. Jim has found that the decisions on the selection of crops, the acreages to plant, or whether to plant at all tend to make farm-

ers line up in the same line, instead of diversifying. But if you don't join the line, you lose access to price supports and may lose your marketing card for the next season—in our case for tobacco; in other parts of the country it's peanuts or whatever. The disadvantages and penalties vary, but they are there.

Some programs are just plain perverse. In the spring of 1992 we applied for the ten-year set-aside program. In order to even be considered for participation in the program, we had to fill out the required paperwork and submit a bid for the amount we would accept per acre to take 160 acres out of production for ten years. That's a tough decision, agreeing not to plant 160 acres for ten years, because a lot can change in that time. But in order to have the chance to decide whether this was what we wanted to do, to even have a shot at playing the game, we had to go through the process. The process required spending a lot of time, often away from an off-farm job, to wait in line, consult with local authorities, and then fill out the necessary paperwork. A couple of months after the initial paper drill, we received word that according to the intricate calculations performed in Washington, D.C., we missed being eligible for the program by a mere six-tenths of a point. Six-tenths of a point! The explanation was equally absurd: if we had shown greater soil erosion on the land that would have been set aside, then we would have made it under the wire. That's to say we had to farm our land less well in order to be eligible for the program. Where is this incentive leading us?

We have always carried crop insurance but fortunately have never filed a claim. By late summer of 1993 our tobacco crop was beautiful, and we were looking forward to a bit of breathing room when we marketed. But a week before harvest a hail storm ripped through and in ten minutes left the crop in tatters. At least it was insured, we thought. We contacted the field office and they sent someone out. He told us that in order to collect we would have to harvest at least two-thirds of the crop; that would provide the evidence that we had planted it! We would have to pay someone to help harvest a crop that we couldn't sell, in order to collect an insurance claim that would barely cover the expenses of putting out and harvesting the tobacco. We couldn't understand the reasoning. But if you are going to play the game, you have to play by their rules. And so it goes.

Over the years we have tried many different strategies to help the farm stay profitable. Sometimes our decisions have paid off, and we've had a profitable year. Other times, what started as a good year ended up poorly with a drought, or too much rain, or hail, or illness in the cattle herd. If we didn't have the income from our off-farm jobs, I don't think we'd still be farming. We couldn't be. There have been too many tough years back-to-

back. But each winter when we look ahead to the coming year, we believe that this will be a better year, and sometimes it is. That hope, occasionally realized, is what keeps us going.

PLANNING FOR BUSINESS SUCCESS

Many farmers would agree with a farmer we interviewed who said that he likes farming a lot, but he just cannot make a living at it. Although the economic pressures on family farms over the last three decades have been economically unsettling, the psychological damage has come from the cycles of boom and bust. When other people fail, it is somewhat comforting to say, "Well, they were simply poor managers." But when your neighbor fails, or when you are reading the bottom line on this year's tax return, you begin to realize that it is not enough simply to work hard and manage well. This situation draws down the wellspring of faith and hope that has sustained many farmers through hard times. A forty-year-old cash grain farmer offered this assessment of the situation: "Well, we always hope that it's going to be better. Personally, I don't think it's going to get a lot better. I know my daddy always told me, all of my life, 'It's going to get better. It's going to get better.' About five years ago he looked at me and said: 'Well son, I ain't going to tell you that no more, because people's been lying to me all of my life, too. It's never going to get any better.'"

As farmers have watched neighbors, good farmers like themselves, go out of business, they have searched for new strategies of action that might lead to greater success. This chapter explores the most commonly used strategies. To respond to the changing marketplace in which they compete, farmers have to adapt their traditional ways of doing business. The seemingly endless search for new ways of doing business adds to the stress of change. The farm business environment is always changing, and just when you think you have hit on a strategy that seems to be succeeding, new factors emerge that alter your conditions of operation or the returns to that strategy. Part of the challenge of success for today's farmers, then, is accepting this uncertainty and handling the stress that comes with a changing business environment.

One factor that compounds the difficulty in identifying and adopting new business approaches is that farmers respond to their changing business environment as both manager and laborer. Farmers are the managers of their businesses, but they are also the ones who do the bulk of the labor, and there is an inherent conflict between these perspectives. Managers should make operational decisions based on the bottom line, what is best for the business as an organization or corporate entity. But doing so may mean adopting strategies that require labor (the farmer) to do more work, learn new skills, or accept less pay. The essential dichotomy between farmer as manager and farmer as labor must affect the decisions about business strategies that are made. In other words, farmers as managers may well make decisions that are not economically rational from the perspective of the farm as a business, but contribute to easing the burden on the farmer as labor.

The farm crisis of the '80s cracked the bedrock of hope and belief in a better future that has sustained farm families for generations. During the '60s and the '70s farmers followed the advice of economists and government officials. They planted "fencerow to fencerow" to build the American agricultural base and America's food security. They modernized their farms, using leading-edge technology to increase yields per acre and growth rates on livestock. They brought more land into crop production, so as to make their new farm investments cost-effective. They specialized, so as to maximize the value of their knowledge, and they participated in every government program that was offered. In fact farmers tried every idea advocated by the experts to improve their economic position. Sometimes such advice did make a difference in the bottom line, but it was not always without costs.

Older farmers remember when debt was a four-letter word, to be avoided at all cost. But for most farmers today, debt is the fuel that runs the farm business. In 1970 national farm debt was $52.8 billion, and 57 cents of every debt dollar was for real estate and the other 43 cents for operating expenses. During the '80s total farm debt averaged $182.7 billion, of which 55 cents was for real estate debt and 45 cents for operating debt. Although the ratio of farm debt to assets nationally rose from 16.2 percent in 1970 to 17.3 percent in 1979, it averaged 18.3 percent throughout the '80s. However, this average masks the fact that a significant number of farmers have much more burdensome debt-to-asset ratios. In 1987 nearly one in four farmers (23.4 percent) had debt-to-asset ratios of more than 70 percent, and 29 percent of all farm borrowers had loans up to their practical limit. What these statistics mean is that one-quarter of all farmers would have to sell 70 percent of everything they have, all their assets, to retire their existing debt.

Debt is still a four-letter word, but now that word is "must." Farmers, mostly the younger ones who started farming during the '70s and '80s, have changed their attitudes toward debt, because circumstances have forced them to do so. You could not start a farm operation in those times, even with family help, without also taking on a lot of debt. Unfortunately, as this farmer explains, prices did not rise to match the increasing debt loads, and the economic scales never seemed to balance: "The last I heard, in the U.S., 60 percent of the farms are debt free. You take the average farmer: he is sixty-some years old and has got everything paid for, so he can take several bucks less for a gallon of milk and still survive. But when you're 80 to 90 percent financed, and I'd say that's low for the other 40 percent, it's different. Some farmers say, 'Well, I don't know why people can't make it. I made it when corn was only a $1.50 a bushel.' 'What did you pay for your farm?' I ask. And they'll say 'I paid $5,000 for it.' But then they sold it the other day for $150,000. Forty years ago you could get $1 a bushel for your corn and make it. But when you tack on inflation, what would we have to get today to be at the same place? Corn would have to be $10 to $15 dollars a bushel, but what we're actually getting is about $2.75 or so. That fellow that hauls feed for Kroger makes more in a year's time than most of the beef cattle farmers around here."

The debt crisis is not just the result of poor financial decision making on the part of farmers. Not all of those who lose their farms do so because they are bad managers, or take excessive risks, or borrow unreasonable amounts. Several factors have led to the growing debt load of farms, but perhaps the most important one is that farmers have to borrow to stay in business. The cost-price squeeze and cash flow shortages have made it very difficult to be able to generate enough cash for operating expenses next year while also paying off this year's debts. So borrowing money becomes the way to float this year's production. During the '70s farmers could leverage more borrowing power, because land prices were skyrocketing and because, on paper, their net worths merited larger loans. Moreover, all the government experts were predicting a strong export market and an ever-growing demand for American agricultural products. The future looked bright. "I'd say that it is worse now than it was in the '50s because you owe so much more money in the '80s than you did back then. Back in the '50s, if you owed somebody $20,000 that was a lot. Now, $200,000 is a drop in the bucket. And think about it, back then the interest rate was 3 percent or 4 percent. What are we talking about now? 13 and 14 percent! Credit got easy in the late '60s and early '70s. You could borrow a million dollars if you wanted it; anybody could. Credit is a good tool, but you've got to know how to use it. I think I've seen a lot of them get caught in the credit system over the past few years. One of their biggest mistakes was to operate on the assumption that land would never go down. Well, it did, and now many of them are paying for it."

Banks and other financial institutions were eager to lend money to farmers in the '70s and early '80s because it looked as if there was no limit to the price of land, and interest rates were rising. Knowledge of the limited returns to farming were overshadowed by the optimism that grew with the rising land values. In other words, the reality of low returns to farming was buffered by the myth of a land market that would continue to rise. This is not too different from the situation of the savings and loan industry in the early '80s. Evidence suggests that regulators were quite aware that many S&Ls had overextended their loan portfolios, having made loans on overvalued properties. Yet regulators did not act; instead, it appears that many S&Ls were encouraged to lend more aggressively in a desperate effort to reestablish a more balanced portfolio. The confidence of farm lenders created the unusual situation of lenders actually telling farmers that they were not borrowing enough, that they needed to borrow more and do more to their farms! "My oldest son bought a farm in 1981, the very worst possible time to buy a farm; the interest was real high and the land was high. He has really had a difficult time trying to maintain the farm. If he knew then what he knows now he would not have bought the farm. At the time they bought it, FmHA [Farmer's Home Administration] encouraged you to do a lot of new work on the farm when he bought it. They spent a whole lot on lagoons and fixed the dairy barn to have feeders rather than milkers. He spent a lot on that work and other things and then had to make payments that have gotten harder and harder to make. The FHA encouraged him to buy big and do all this extra work on the place, and then they turn right around and make it real hard to keep up the payments."

Many farmers, especially those who entered farming in the '70s and '80s or those who expanded their operations at that time, found themselves literally swimming in debt. It was a debt load that might have been manageable if the weather had been right, if farm prices had remained stable, if interest rates had stayed low, and if land had continued to rise in value. But the droughts came, interest rates continued to rise, the grain embargo began, the dollar became less competitive on the international market, reducing exports even more, and the bottom fell out of the land market. Suddenly the debt was not manageable any more, and many farmers faced the very real prospect of losing their businesses. The fact that so many farmers—so many of the best farmers—were put in this position suggests to those who survived those years that it cannot all be blamed on bad business decisions. This couple is just like many who left farming during the '80s; here is his description of the late '70s and early '80s: "The worst part was when they had the high interest rates. I'd just bought my daddy's farm. Then I bought two other farms, but it was when land prices had pretty well peaked. Right after interest rates zoomed to 18 percent, we had bad weather and prices started to go down. In a couple of years I wasn't hardly making enough money to pay the interest. A lot of years we had to borrow the money to pay the 18 percent interest, and all that did was put you deeper in debt. The next year instead of having this amount of interest, you've got the extra amount of interest on what you borrowed to pay the interest on the money you had to borrow the year before. It seemed we just kept getting deeper and deeper in debt. Then tobacco got up to about two dollars a pound, and it looked like we would finally have a chance to get the debt down; at least we could keep the interest paid. Then the government cut the tobacco support prices fifty cents a pound and cut the marketing quotas. That was almost like your crop had been cut in half; it cut my tobacco income in half. It was quite a blow. I thought we would have to get out of farming, but somehow we've hung on."

The combination of the cost-price squeeze and rising indebtedness means that many farmers, especially the younger ones, see only a long dark tunnel ahead, with only a small hope of "settling up" by the time they retire. Unlike many of their parents, who entered their fifties with their farms and equipment all paid for and a chance to begin saving, these younger farmers see only a lifetime of debt. "We might have a forty-thou-sand-dollar tractor sitting out there, but you owe thirty thousand dollars on it. With most people our age, their midthirties, who are farming, it is not all ours, and it will be a long time before it will be all ours. By the time we get the tractor paid for it is time to get a new one. With farm prices where they are and all of the interest that you owe on your farm mortgage and operating loans, all you can hope for is that you have some appreciation in land values so you can recoup some of these costs. The way it is now, we will be lucky to recoup the purchase price of the farm by the time we're ready to retire."

The high debt-to-asset ratios (60 percent and more) that a growing number of farmers carry are far less common in nonfarm businesses. Most nonfarm businesses carry debt-to-asset ratios less than 50 percent. Typi-cally, nonfarm debt loads higher than that would lead to either the lender's withdrawal of support or the owner's decision to get out of the business. But farmers with these debt-to-asset ratios are encouraged to stick with the enterprise. Lenders see little opportunity to recover their investments by forcing bankruptcy, and they share the farmers' optimistic view that things will get better. Furthermore, government interventions (for example, emergency loan programs) often buffer the accumulating effects of high indebtedness, enabling farmers to continue eking out an existence on the farm. And so, debt becomes a familiar partner in the farm enterprise, and debt loads that would drive any other business owner to cease operating are considered "normal" costs of doing farm business.

Like any other business owner faced with a financial crisis, farmers have looked for and tried different tactics to improve their competitiveness and give themselves a financial edge. By and large, the solutions that farmers have tried to improve their competitiveness are the same used by most other businesses: expanding the business, mechanizing, specializing, and diversifying. Pursuing any of these solutions has often meant increasing indebtedness, so they are both a cause and a result of the financial crisis facing farmers.

For many farm businesses the answer to the cost-price squeeze and high debt loads has been to expand the operation. There is a logic here. In other businesses, if you are trapped in a cost-price squeeze (as the steel industry was in the late '70s), you either go out of business or reduce the squeeze by reducing costs or getting higher prices for your product. For farmers, the option of getting higher prices for your product by improving its quality or penetrating new markets first is limited, so they seek the first option, reducing costs. Reducing their costs often requires expanding their production, hopefully without substantially increasing operating costs.

There are several ways to expand the business. You can produce more by cultivating more land, typically rented land so that you do not have to bear the purchase costs, or you can raise more livestock and hope that by simply selling more, even at low prices, you can make enough to get ahead. In other words, you maintain the same margin of profit, but it is on a larger base of production. Or, you can continue to operate on the same scale but increase productivity by mechanizing or reducing the costs of what you produce. Here, you continue to produce the same as before, but you lower the cost of doing it, and therefore your margin of profit increases. One way to reduce the costs of production is by cutting back on hired labor and putting more of your own or your family's labor into the operation.

But these options are not without their own costs. More land, better equipment, more livestock, the seed for more production all cost money, and that means borrowing more. Thus, the farmers who try to overcome the cost-price squeeze by expanding production often find themselves on a treadmill: in order to make more money by expanding production, they must increase their investments in the farm, often by taking on more debt, which in turn increases their risks.

During the '70s and the '80s government officials, lenders, and all the pundits told farmers that the solution to their problems was to get bigger. Foreign markets for American agricultural products seemed unlimited, and farmers were told that it was their patriotic duty to make "agripower and the agridollar" a part of America's foreign policies. Bigger was better, and bigger was the solution to both farmers' problems and America's growing export deficit.

Getting bigger might have been a solution at other times and for some farmers, but in the '70s and early '80s this solution required paying more for land than at any other time in history. Few if any farmers had the cash resources to buy land selling for fifteen hundred to two thousand dollars an acre, and so they borrowed. They borrowed against the land they already owned, or they borrowed on the basis of expected returns. They borrowed, bought the land, and then the bottom fell out. It became clear that two-thousand-dollar-per-acre farmland could only be paid for by selling it for some other use. Expensive farm land simply cannot produce enough cash flow to pay for itself, at least not by farming it. "When land got to twenty-five hundred to three thousand dollars an acre, everyone thought that there was no limit to how high the land was going to go. My daddy said that land may get higher but he couldn't see paying more for something than it can produce to pay for itself. He said you can only make maybe three hundred dollars an acre of corn and that means you can't farm your way out of fifteen-hundred-dollar-an-acre land prices. Me as a farmer, I don't want to see land high because I'm not in the business of selling land. I don't care what the bank says my land is worth on paper. I like for land to be cheaper, where I could buy it and be able to pay for it by farming it."

Another way to expand production is to reduce labor costs by mechanizing. Labor is one of the largest production costs for farmers, as it is for other businesses. And, like other business owners, farmers have attempted to control this rising and variable cost of operation by mechanizing, replacing labor with machines. It is a solution that is increasingly available to all types of farm operations.

A grain farm owner: "That is how we're facing the labor problem is with machinery. It's not the fact that at the beginning, machinery is cheaper, because it isn't. But you don't have much choice, because on our farm of twenty-nine hundred acres, we used to have ten men and now we operate with three men and the machines."

A dairy farm owner: "We used to have two to help milk. So we bought automatic takeoffs so that one person could milk by themselves. We figured we could pay for them from the salary of two people. So that is what we did and it has worked a whole lot better ever since then. Now my husband can do all the milking by himself."

But machinery costs money, and more every year. In the late '50s a gas tractor with two plows and three other pieces of equipment could be had for twenty-six hundred dollars, but today a similar machine would cost nearly twenty-five thousand dollars. You end up having to buy machinery as you do land, with borrowed money, adding to the debt load. In a sense, expanding production by buying more land or mechanizing is a paradoxical solution: you can produce more and at lower short-term costs, but you also end up increasing your long-term debt and your interest payments.

Not everyone heeded the experts' call to "get bigger!" Many farmers felt that the way to financial success was to "stay the course." They were committed to continuing doing what they had always done, on the same scale, and believed that over time, costs and income would even out. This option often means working harder—more hours doing work that other farmers do with bigger equipment. The farm is a greedy business, and you pay the costs either with cash or with your own labor. The comments of the following two farmers, who operate very different-sized farms, illustrate this rule:

"I still do things pretty well the old-fashioned way in that I don't have more horsepower than I need in tractors, and I don't have more width of disk than I need. I just figure I can spend more hours on a tractor seat and do it that way, cause like I said, I bought this eighty-seven acres in 1979, and not too long thereafter it became pretty obvious at least at that time, it wasn't worth the price I paid. A lot of people would probably go crazy trying to handle ten acres of tobacco with the machinery I do, but I spend a lot of my own time. The hogs and tobacco take a lot of time, and I've got more time than money still, so that's pretty well the way I operate."

"We're still very small, only five hundred acres. In fact, we've had people wonder how we can make it farming without one of us working off the farm or without farming more acres than we do farm and still live as comfortable as we do and have the equipment that we have. But we learned very early that if we were going to stay in the farming business we had to set our priorities and know what we had to have. We only buy new machinery when the old can no longer be repaired, and we tend to buy equipment a bit smaller than most others would because our time is cheaper than the equipment."

Government regulations have often influenced the decision whether to acquire new technologies in order to expand production. Especially in the dairy business, meeting government regulations and staying competitive has meant making substantial investments in new equipment (for example, refrigerated holding tanks and automatic take-offs) and in improvements in buildings (for example, concrete floors and permanent plumbing fixtures). Over the last twenty years in dairying, farmers' strategies for success have been defined by government regulations with regard to sanitation, safety, and health. This farmer points to the empty choice he faced. He just did not have the capital to modernize his dairy farm. "We had a small milking parlor; we milked two at a time. It was either get out or get bigger. Bigger meant spending forty thousand dollars for a new parlor, and we just didn't have that kind of money and we couldn't get it." To stay in dairying would have required a major financial investment, one that this farmer could not afford, and a debt he could not foresee being able to pay off, even if he could borrow the money. So he quit dairying and switched to beef cattle and corn.

In other businesses a strategy for improving competitiveness is specialization, selecting a market niche and producing a better product at a better price than anyone else. Farmers can specialize, too. Specialization provides an opportunity to refine and focus their technical knowledge and skills as well as their management skills. But unlike other business operators, who can give a product a brand name and market it as different from or better than anyone else's similar product, a farmer cannot market John Smith's corn or Bill Watson's soybeans. Specialization for a farmer simply involves focusing his or her energy, knowledge, and equipment on one commodity. It is a strategy that is common in cash grain crops (for example, wheat, corn, and soybeans) and in the dairy industry. "It's become more specialized. You used to just try to do a little of all of it. You raised hogs, beef, had a garden, milked cows, raised chickens. Now we have farmers that just dairy, or raise beef cows. That is all they do, have a dairy and maybe grow the crops to feed their animals. You don't have as much generalized farming any more. Raising a 2,000-pound crop of tobacco was considered a big crop of tobacco for a family before. But now, we'd grow that on about two acres. Today we have people that will grow twenty acres or more of tobacco; that's 100 to 150,000 pounds of tobacco. Today's farm is not just a little of this or a little of that, hit or miss type of thing. They put a lot of thought and work into that one thing. But if they fail, they fail to survive."

If you are going to specialize it helps to start with the right inputs. For example, if you are going to specialize in grains, it helps to have the right kind of land—land that is rich and flat. "I had bought a farm and I put all new steel posts, treated corner posts and barbed wire up, and I built a new barn. I had registered beef cattle then. I had just got the operation going well, and I took a big steer to market in 1965 weighing twelve to fifteen hundred pounds, and I got fifteen cents a pound. I said, 'That's it.' I came home and that winter I sold the cattle. I pulled all the fences up and started raising winter wheat and soybeans. The neighbors thought that I had gone crazy. But that first year I more than doubled my net income, and by getting rid of the cattle I had a lot less work. See, I've got good ground, no rocks; it makes good soybeans. Last year we ran fifty bushels of beans an acre. You can't raise cattle on pasture on that type of ground and earn anything near what I do with my soybeans."

Specialization is a strategy that allows farmers to buy only certain kinds of equipment and refine their knowledge in one area. It is a strategy that can be profitable. Simply consider Iowa corn farmers or the wheat farmers of Kansas and the Dakotas. But specialization is a risky proposition, as many business owners know. If the demand for your product declines or something disrupts your production, you have nothing to fall back on. And farmers recognize this risk, for as one noted: "That's one of the reasons that farmers have gotten in such a bad shape in the last few years, they specialize." For farmers who have specialized, if they experience one or two bad years—from drought or excess rain or early freezes or low prices—they need to increase their borrowing simply to survive, increasing their indebtedness and making it ever harder to get ahead. Once again there is a trade-off. The increased productivity and efficiency that comes with a specialized operation versus the higher risk that comes with hanging all your income on one crop. "The people that have gone broke farming are the ones that have stuck strictly to row cropping. Farming right now is going back to like it used to be years ago when you had to do just a little bit of everything. You cannot stick to one primary income—just crops, just livestock. You've got to be able to do several things, because if one don't hit one year maybe the other one will."

Although it might be a good idea to diversify your operation, it is not always easy to do. For example, during the '70s, when farmers were urged to plant fence row to fence row, many tore up their fences and transformed their diversified operations into single-crop enterprises. To add livestock back into your operation means having to reset fences, or build farrowing barns for hogs, or make any number of other changes that entail spending more money. Money is just what many farmers do not have, and taking on more debt is just what they do not need. What happened in Grain County during the '70s illustrates this process. Several farmers there commented on the fact that during the '70s many stopped raising tobacco, shifting instead to only cash grains. But times made this a risky decision.

Tobacco may be labor intensive and hard, sweaty work, but it has one of the highest returns per acre of all crops. Today, many see tobacco as an integral part of a diversified farm operation. "You've got to get your foot in a lot of different things to be able to have different incomes at different times. If something fails you've got to have something else to fall back on. I'd like to have everything, I even wish I had chickens and turkeys. I don't like to have all of my eggs in one basket. You've got so many elements to fight if you want to succeed as a farmer. I'd rather have four or five different crops and if one fails, then fall back on another one. Tobacco has been real good to us, and it helped save a lot of farmers when grain prices began to go down."

Just as many other businesses have come to recognize the financial risks inherent in specialization, so too many farmers have come to see the advantages of the way farming used to be. Of course, some farmers never did specialize, they began diversified and they stayed that way, and others have come back. The following description gives a good sense of what is entailed in a diversified farm operation: "I guess the main thing is that with the diversity of our operation we have enough cash flow to be able to take care of the day-to-day needs. The only thing we have to borrow money for are the major purchases. This year's soybean crop has given us enough to sell to pretty much pay for all this year's operating costs. Even if the beans don't make it, it is still going to be a good year. The wheat was a good crop; so was the barley. Hogs are still at a fairly decent price. Our tobacco looks pretty good. We've got enough feed in the bin to take care of the cattle for the year; even if I have to start feeding early I won't have to buy any. It seems like when one thing around here flops or doesn't come up to what we think it ought to, something else kind of picks up the slack. We are able to go on."

Some have argued that a diversified farm operation prevents the farmer from developing the knowledge base essential to maximizing efficiency, or that the demands of one aspect of the farm business prevent her or him from doing well in other parts. But those who operate diversified farms have a different perspective on the trade-offs involved. "I see diversification as an opportunity to get into several things that complement each other rather than take away from each other. The alfalfa goes with the beef because I salvage a lot of my nonmarket alfalfa through the beef operation. We just have enough grain to play the government programs to the maximum income from that. But you see, our land is rolling and it never was suited to grain production to start with. So part of it is protection from risk and the other part is trying to do a better job managing the land."

The clear advantage of a diverse farm operation is that there are multiple sources of income. If one part of the business does not do well, there is still the likelihood of making money this year from another part of the enterprise. Moreover, with diversity, the farm income is likely to be spread more evenly across the year, reducing the knotty problem of cash flow shortages. Finally, multifamily operations can maximize the returns to the expertise of the individual partners as each one specializes in making more profitable his or her part of the diversified operation. Farmers adopting the diversification strategy say they have survived the hard times and sometimes even paid off debt because they were diversified. "We've always been diversified because as a rule something would make money every year. When grain prices dropped we still had our cattle, and cattle prices were good. We fed our grain to the cattle, and for three years the livestock was the only thing that turned a profit. I guess you could say that even with low prices for grain, we fared better than most being diversified and have even been able to pay off some debt."

However, diversification has its disadvantages. To maximize your returns, you need a system of crops and livestock that uses your labor and equipment in a sequential manner. That is, you must plan so that your hogs are not farrowing when you also have to be setting tobacco, or your winter wheat does not have to be harvested when your hay is also ready to be baled. Having a diversified operation also means that the work is never done. "Being diversified means you just don't ever get caught up. There is always something to do. Just as soon as he's finishing the double crop beans then you've got to get to the tobacco patch. Just as soon as we get that cut in August, I've seen him go from cutting the tobacco today to harvesting with the combine tomorrow with no break. He does the majority of the feed grinding after supper at night. Back last fall our neighbors could hear his feed grinder going every night because that was the only time he had to grind feed for the cattle was at night."

One type of diversification that many experts are suggesting to farmers to improve their financial situation is to shift into alternative commodities, that is, growing crops or raising livestock that typically are not raised in the farm's geographical area but for which there is domestic demand. Alternative crops can include shitake mushrooms and elephant ear garlic, as well as more common crops such as tomatoes, peppers, cucumbers, and strawberries. Alternative livestock might include sheep, rabbits, ostriches, and llamas. But shifting to alternative crops or livestock is not as simple as it sounds, just as it would not be simple for manufacturers to add a new product line to their production operation.

Diversifying into alternative commodities always requires finding new markets and acquiring new knowledge about how to raise the new crop or type of livestock successfully. Land will also need to be shifted away from a traditional, known crop into the production of an unknown commodity. It requires additional labor to fit the crop into the production schedule; and it may mean buying new equipment for planting or harvesting. For these reasons, switching to alternative commodities is the most risky of the diversification strategies. In Kentucky, vegetables are the most common alternative crops, and the two key problems that always emerge in discussions of these alternatives are labor and markets.

The labor problem is based on the fact that on a small scale of production, most vegetables must be picked by hand. If a farmer is having trouble finding and affording the labor to bring in the tobacco crop, an activity that is concentrated within a two-week period, the problem of finding and affording labor to harvest pickles or cucumbers for a month or so is even worse. As a result, those who raise vegetables tend to rely upon their family members for the bulk of the work, and without enough family labor, these crops just are not very profitable in terms of labor costs. "Crops like pickles are for when you have lots of family. I've got a nephew here that has five or six little boys and all of them out there helping harvest every day. I talked to him last year. He said with seven of them they could pick up forty-five to seventy dollars a day selling their pickles and they wouldn't have any other way to do this. He said that helps them make a better living than without the pickles. But most people don't have five or six kids, so if you had to go out and hire that labor, even paying minimum wage, you just can't afford that kind of cost. See for my nephew, by not having the labor costs, that's where his profit came."

Having enough hands to harvest is important, but more important is harvesting at the right time so that the crop brings the maximum return at the packing sheds. With tongue in cheek, this woman farmer explains the "Russian roulette" of successfully growing vegetables: "Red peppers are fine. They are a great thing to raise because for one thing, it is easy to pick them. You don't have to make a decision as to are they big enough, are they ripe yet? If they are red, they are ready. They are not going to grow after that. If you cannot really pick them today, you can still go out tomorrow and get them. They do not start out as gooseberries and if you miss them, the next morning you find out that somebody with a bicycle pump pumped them into melons! That happens with pickles. Pickles are sticky and you have to crawl around them and have to always look at the darn things and wonder if they are the right size. Then they grade them. They have to be certain sizes; the smaller the pickle the more the money you get for them. And it always seems that only a few of them are the very best grade. There you are. They spoil so fast. I have nothing against pickles. I even eat them on occasion. They are not bad. But if you are working hard, you need something that is a little bit more certain."

Marketing alternative crops, as noted by the last farmer quoted, is even more critical. You need knowledge of what the market is looking for and access to a market where your time and labor can be transformed into profits. Although there are livestock, grain, and tobacco markets in nearly every county in Kentucky, there are very few packing houses where a farmer can take vegetables to sell. Several farmers noted they had to drive fifty or sixty miles one way in order to get their vegetables to a market. So the time it takes to get your vegetables to market may become a factor in their profitability. "It has never been good policy, I think, to encourage somebody to raise something and forget about the market until the stuff is ripe. Then what are you going to do with it? You've got to have a market. There is not a food processing plant within a radius of one hundred miles. Peppers were fine, but the trouble was that you had to go forty to fifty miles real early in the morning in order to sell. All of these pepper farmers are standing in line with those trucks. It took half a day out of my husband's farm life twice a week, maybe a whole day, because he would have to hang out there forever waiting for them to grade those things."

Another problem with vegetables, from the perspective of farmers, is that the price you receive is based on how the graders at the packing shed evaluate your crop. The price, in other words, is not set but variable. With cash grains and livestock, you have a good sense of what your product will bring at market. There are not many factors that would substantially lower the price of your crop below the current market price. But when you leave the farm with a load of vegetables for the packing house, what you actually sell and the price you actually receive may be substantially different from what you expected to get when you left home that morning. "We tried pimento peppers one year. They have to be a certain size or they will not buy them. When the pepper goes to the factory there is a certain machine they run them through. If they are too small, they won't buy them. Lots of times we'd end up coming home with half of the peppers we'd taken to market."

Finally, through word of mouth and the media, it is clear that marijuana is a part of the informal farm economy in some places. Throughout Kentucky historical markers indicate that hemp was once a valuable agricultural product. Today it keeps some farms afloat and remains important in some communities. The cash from marijuana sales finds its way into local stores and businesses, even though no taxes are paid on this income. Unofficial estimates are that marijuana is the leading cash crop in Kentucky, and although no one admits to raising it, many know of someone who knows someone who does. In the United States marijuana clearly is being raised as a recreational drug, but in Canada, low THC (the active hallucinogenic ingredient) hemp is being raised in selected areas for diverse agricultural uses—as livestock feed, as fuel, and as an additive to cotton for higher quality cloth—and for export to a growing world market.

Depending on how it is done, diversification can be as risky as specialization. There are added investments to be made, new skills and knowledge to acquire, and new markets to find and understand. Yet many farmers enjoy the opportunity to try new crops, new technologies. There is an excitement that comes from being entrepreneurs and the challenge of change, though it can also be stressful because of the risks inherent in change.

Since the Great Depression and the Dust Bowl years the federal government has become an increasingly important "player" in the agricultural economy. Government programs run the gamut from low interest loans to crop insurance to price supports to production controls to soil conservation to managing environmental quality. From the government's perspective, these programs have many purposes, some stated as clear policy outcomes, others as implicit but intended consequences. Policy objectives include providing a more stable market for farm commodities, maintaining the quality of the natural environment, and "saving the family farm."

An implicit consequence of most farm programs is to establish and maintain low food prices, and America does have the lowest-cost food of industrialized nations. Government officials often deny, however, that maintaining low food prices for consumers is a conscious purpose behind any commodity program. Farm families become frustrated with the situation because they believe that unless this purpose of commodity programs is made explicit, the public will assume that they are simply another government "give-away" to a special interest group.

From the perspective of farmers, the overriding purpose of government farm programs is to stabilize farm income. Farmers participate in government programs as another strategy for improving their competitive position. But farm programs are designed in Washington, D.C., far from the fields of Kentucky, Iowa, or Nebraska, and because they are national programs, they treat all farms as if they were alike and assume that the conditions of farming are like the conditions under which all businesses operate. The flaw in this approach is revealed in the following comment, which illustrates the tensions that emerge when farmers and government try to manage farming together. "The government doesn't realize the importance of timing. Farming is about the essence of time. They eventually released the set-aside fields during the drought, but it was too late. We needed to be able to cut it on Friday but they didn't release it until Monday. Three or four days may not sound like much to them but it makes a big difference to us. Last year a lot of cover wheat was sowed on set-aside ground, and they released it earlier so we could cut it for hay. But this year they waited until the wheat was too far along to be useful as hay. I told the supervisor that they did that on purpose, and he just kind of looked at me and grinned."

This farmer's interpretation of the intent of the delay in releasing set-aside fields for haying during the drought may be wrong, but his point is well taken. Federal farm programs are managed at the national level, and most of the administrators have never been on a farm and have never farmed. They establish eligibility requirements and procedures for implementing programs that apply nationally. But farming is a business affected by local conditions—rainfalls, hail storms, and insect infestation—and the few days or weeks it takes for Washington to make a decision about relaxing program guidelines can literally mean the difference between survival and financial disaster. In this sense participation in farm programs can actually increase the economic risks confronting farmers.

Farm programs are political creations. Policies and programs change with administrations, and entire programs can be eliminated or changed as a result of interest group pressures. Farm programs often become items for trade in budget negotiations, with their eligibility requirements and funding traded for votes on other items. Farmers resent seeing an increasingly important factor in their economic success treated as a bargaining chip. Yet, although farmers tend to be some of the most vocal political supporters of smaller government, at a practical level they also see a need for farm support programs. This is one of the ironic contradictions between beliefs and practice that has emerged with the changing conditions of agriculture today.

Federal farm policies and programs are alternately cursed and praised by farmers. Farmers curse what they feel is an infringement on their right to make independent business decisions, but they also believe that under current market conditions they cannot survive without government intervention. A farmer's management decisions are not enough to overcome the crippling effects of the cycles of boom and bust; they require assistance at a higher level—the kind of help only governments can provide. Yet many feel that accepting that help is like entering a pact with the devil.

To understand the mixed blessings of farm programs, the sense of frustration and deliverance that farm families feel when the subject of farm
programs comes up, it is necessary to see how these programs work from
the perspective of farm families. Three examples of widely used farm programs will illustrate the difficulties.

The dairy herd buyout, begun in 1985, was designed to reduce the
number of operating dairy herds in America, in order to lower milk production and stabilize milk prices. Dairy farmers agreed to sell their herds
for a price guaranteed by the government and also agreed not to reestablish their herds for a specific period of time. A logical question that a
dairy farmer might ask when considering participating in the program is
how the income would be treated for tax purposes. A logical question,
but apparently not one anticipated when the program was developed, as
one farmer who is also a CPA found out when he was trying to help a
neighbor decide what to do. "This farmer I was helping was thinking about
participating in the dairy buyout program. He went to the ASCS [Agri-
cultural Stabilization and Conservation Service] office to find out what
the tax consequences of this were going to be. They couldn't tell him.
Here he was, this little small guy out there, trying to make a decision
based on incomplete information on what the federal government said it
was trying to do to help him. I went round and round trying to find information for this guy from the local ASCS office and the offices above
them. Nobody could tell me how this income was going to be treated.
How was he supposed to make an important decision like this without
knowing how he was going to end up economically if he participated? It
was a terrible situation."

Something else happened during the dairy herd reduction program
that seems to happen with other farm programs—the rules changed in
the middle of the game. In this case dairy farmers had been told that they
could rebuild their herds after ten years, and this provision was a factor
in many farmers' decisions to participate. But then the period of no return was extended by several more years after they had entered the program, thereby changing the whole basis for farmers' calculations of the
relative costs and benefits of participating in the dairy buyout program.
"We had it all figured up, what it was going to cost us to produce the milk
and what they were going to pay us for not producing it. We thought it
was pretty simple. But then they changed things and that is the way it is
with the government. It takes a Philadelphia lawyer to read all of the fine
print. Any contract you sign with the government, if they tell you one
thing there is another place that seems to say that they can do the opposite too. Like the dairy buyout program, six months after it went into ef-

fect, they changed it. So they never told you in the beginning that you weren't going to be able to rebuild your herd after a period of time."

Other farmers also deal with the problem of changing rules. Many farm programs have a price support function, that is, there is a minimal price per unit guaranteed by the government. If the market price drops below this level, the government will buy commodities at the promised price, thus guaranteeing a minimum return to farm businesses for their production. In most programs, the costs are offset by a per unit fee assessed on all commodities sold, so that in effect farmers themselves pay for the government purchases. Not all price support program are for food commodities (tobacco and cotton are included, for example), and not all commodities are associated with price support programs.

One consequence of price support programs is that the market price often ends up being only a few cents above the price support, and so, in effect, the government price support sets the market price. The commodity price support established by the government at the beginning of each year, then, can be considered by farmers as they make production decisions, such as how many acres to plant or how many acres to set aside. The problem is that what starts the year as the price support may change six months later in the middle of the growing season. This happened to tobacco farmers one year, when the price support dropped nearly twenty cents between January and July. Suddenly the January planting decisions designed to maximize returns while minimizing costs were worth less than the paper they had been written on. Lease prices already paid to market under someone else's allotment, which had been modest given the price support and the expected market price, were now too high. Others have experienced the same yo-yo decision making from Washington with equally serious financial consequences. "This year's corn crop started last September to August at $2.92 a bushel. But Monday they dropped that $.40 and we've got to refund the difference to the government so the program won't cost the government anything. You see, the government's more or less guaranteed that we would only get $3.00 a bushel for corn, and we've got to have that to break even. If the price of corn does go to $3.00 a bushel, then the government won't be out a penny."

Another farm couple note that participation in a farm program is no sure guarantee of income as promised. "The last three years we have been in the grain set-aside program. But they have been slow about paying us what they owe us. In fact, they are two years behind on most of it and not paying any interest." He went on to note that if *he* were two years behind in paying his income taxes, he would surely have to pay interest *plus* a penalty. But as a participant in a farm program all he can do is wait.

A major problem facing any farm family considering entering a federal farm program is trying to figure out the eligibility and participation requirements. As in all federal programs, there is a lot of legalese framing the purpose and functioning of the proposed programs, and wading through it requires, as one farmer said earlier, a "Philadelphia lawyer." The purpose of county Agricultural Stabilization and Conservation Service (ASCS) and Soil Conservation Service (SCS) offices is to provide local farmers with information, to interpret the federal programs, and to assist farmers in completing the forms required for participation. But despite the best intentions, local representatives are sometimes as unsure of program requirements as those they are assisting. The sheer complexity of the programs, combined with the nearly constant "tinkering" with them by Congress and the U.S. Department of Agriculture, means that local offices need to relearn the system every year—and sometimes more often. Moreover, the massive amount of paperwork required of farmers to manage these programs can overload both the system and its participants.

One of the most difficult requirements for everyone involved in any kind of crop program is the need to be absolutely precise as to how much land is in what type of production. This information is used to determine eligibility for crop set-aside, Payment-in-Kind (PIK), and a host of other grain programs. But it is not easy to reach the level of specificity that is now demanded. "We now have to go to the ASCS office every time we turn around. We have to report, report, report; every tenth acre of land has to be reported. You have to tell what it is doing, what you are using it for. The other day I was turning in a crop report on the one farm and it was one-tenth acre off. I've never seen paperwork like this in all my years of being in farm programs. You wouldn't believe the forms I had to fill out for one-tenth an acre of land! When you've got six hundred acres it's easy to be a few acres off on every crop."

Trying to deal with these programs reminds some farmers of Alice in Wonderland's meeting with the Cheshire Cat, who told her that when he used a word it meant exactly what he wanted it to mean, nothing more and nothing less. The following story of one farm family's encounter with the bureaucracy may be an extreme example, but it is not unusual. "I signed up in January for the PIK program. That is when they give you grain certificates to purchase grain from government surpluses instead of you raising your own. It was right at the beginning of the sign-up. It was the day after the deadline before they got ours straightened out. I said, 'Well, what do I need to do?' The supervisor finally said, 'Well, I can't tell you. You write down the way you want to do it and submit it to the board for consideration. If it is feasible, then we'll do it that way. If not, we'll come back to you and you'll have to redo your proposal.' I said, 'Why don't you just tell me what I am allowed to do and then we'll go that way.' He said, 'I can't do that.' I said, 'Why not?' He said, 'We're not allowed to help a farmer get any more payments than he already is. We're not allowed to help. You can maybe get you a lawyer and he might be able to help you.' I was about ready to pull out all the hair that I have left.

"So this year, we started early again. It was the day after the deadline before they ever signed the agreement. That delay cost us quite a bit of money. After you signed up you would get your advance, and part of your advance was in grain certificates. We wanted our certificates to roll our corn over, but we didn't get our certificates on time, so the margin dropped in corn. We lost because the price of the corn went up and we couldn't buy as much of the government surplus because we had to pay a higher price. That delay cost us several thousands of dollars, which didn't set too well with me since I had submitted all the required paperwork on the opening day. I was one of the first ones that brought them into the ASCS office and I was the last one to get them signed. There are real nice people that work at our ASCS office. But a lot of the stuff they don't know because nobody tells them. If a problem comes up, they'll send it in to the regional office and try to figure it out. It might be two or three months before they learn something. I asked someone over there: 'When are you going to get to ours?' She said, 'Well, I'll tell you. I got to it Friday and it was so much trouble that I put it on the bottom of the stack. I'll probably get to it next week.' Just on those certificates we lost ten thousand dollars."

Because this farmer and his father have formed a corporation as part of their formal agreement, their participation in the PIK program is evaluated under different guidelines than for other farms, further complicating an already intricate web of eligibility requirements and exclusions, with enormous demands for precise record keeping on the part of the participating farmer and the need to confirm the use of every tenth of an acre by field office personnel. The considerable amount of administrative bookkeeping required by government programs is one reason why large farm enterprises are more likely to participate than smaller operations. Large farm operators have more to gain from these management tasks than farmers with smaller enterprises, who see a lot of hassles for little return. As a result, the bulk of farm payments (about 7 in 10 dollars) go to only one in three farmers.

Is it any wonder, then, that farmers and the workers in the ASCS and SCS offices feel alienated and often angered by the process? Farmers realize that office workers are doing the best they can with the confusing information they receive from Washington, but they cannot yell at those in Washington who design what seem to be intentionally confusing programs. The field staffs realize that the farmers are not really mad at them. After all, they are neighbors and often share farm work with each other, but still, it is hard not to react.

Another set of programs that lead to great frustration are those designed to encourage soil conservation and more successful management of natural resources. After the destructive effects of the Dust Bowl years, the federal government became a major supporter of wise land and resource management practices, providing technical and financial assistance to farmers. But today many farmers feel that farm programs have conflicting effects. Sometimes they encourage production on land ill suited to row crops and at other times encourage participation in set-aside programs that idle productive land for extensive periods. Because of changing environmental concerns, some farmers have been told that they can no longer use land now defined as wetlands, while others, who need to develop new water sources for livestock or irrigation, encounter regulations that hike the costs for those projects. Although federal programs will provide financial assistance for conservation projects, it is not without some cost to the farmer, as described by this farmer: "First, everything you get from the government has to be done according to their regulations, even if those regulations don't really make sense for your land. Then, if they are going to pay 80 percent for the digging of a pond or a small lake, you have to pay income tax on all that money. Not only that, but that money is counted as income and puts you into a higher tax bracket so you end up paying a higher rate on all your income. So a lot of times I'm not sure that you gain much financially from participating in these conservation programs, and they don't always make that much difference in terms of making the land better."

One of the most common myths about farm programs is that they are a form of "give-away," a welfare program for farmers. This farmer points out correctly that everyone pays taxes on any income they derive from participation in farm programs, and many require cost sharing as a condition of participation. Still other programs entail indirect cost sharing through fees or assessments on marketed commodities, which are collected either by the government or by the commodity associations, who then use these funds to subsidize the government programs. Yet even farmers often complain that farm programs build too much dependence on the government and alter the free market for agricultural products. But the phrase "welfare program for farmers" generates considerable hostility. From their perspective there is a difference between welfare, which is given to those who do nothing to earn it, and farm programs, which are more like low interest home mortgages to veterans who have earned this assistance for their contributions to society.

Most farmers recognize that some level of participation in government programs is essential for their financial survival. A considerable proportion of our domestic grain production is sold on the world market. In the early '70s the federal government recognized that America's farm products represented a consistently strong seller on the world market and encouraged farmers to produce more to build America's "agripower." During the last two decades the export of farm commodities has helped reduce our trade deficit, and the impact of farm sales on our balance of trade is another reason for Washington's support of farm commodity programs.

The concept of "agripower" has often meant using the sale of U.S. farm products as a tool to reward or punish other nations. Although the effectiveness of grain embargoes as a tool of foreign policy remains a point of debate, there is no question as to their consequences for grain farmers who suddenly find themselves locked out of some of the largest world markets. When grain embargoes are imposed, commodity prices plunge, and a crop that once appeared to be able to generate enough income to repay some long-standing debts may become worth more simply plowed under. The farm commodity programs act as a counterbalance to the vagaries of the world market, yet sometimes foreign policy diminishes their effectiveness. This, then, is another example of the "good news–bad news" that feeds the love-hate attitude toward farm programs and the government's involvement in farming, as illustrated in this comment: "Well, the government programs have been a lifesaver for a lot of farmers, for myself as one. I think they were good. But then again, I think that the government had a whole lot to do with the surpluses and the problems that we had leading up to the farm recession. Yet I think that the farm programs were timely, and I think they were necessary."

Farmers also recognize that they compete in a world market where the playing field is not level and the rules are not the same for all the players. The U.S. government has strongly resisted any outright subsidies for farm products, demanding that American farmers compete in a "free" market, but many other governments have no such concerns. In a sense European and other governments recognize the trade value of their farm products and do what they can to improve the competitiveness of their farmers on the world market. "We have to meet foreign competition. England pays its farmers four dollars a bushel plus whatever they can get on the world market. How can we compete with them on the world market if they got four dollars to start with? France and Germany does the same thing."

New uncertainties have entered the equation of agricultural exports. The North American Free Trade Agreement (NAFTA) and the General Agreement on Trade and Tariffs (GATT) will alter the kinds and levels of support national governments can provide agriculture and will reduce or eliminate agricultural trade barriers. At least this is what is said. But farmers and agricultural analysts are not sure how these new international agreements will affect farmers' opportunities to sell their commodities on the international market or the prices they will receive. Although NAFTA and GATT will affect the foreign market for all businesses, other businesses can shift to foreign locations in search of lower labor costs or other competitive advantages in order to maximize the potential advantages of these agreements. But for farmers, all the new agreements mean is a new uncertainty added to the business of farming.

From the perspective of farmers, there are inherent contradictions in praising free market competition, using America's farm products as tools of foreign policy, supporting farm commodity programs, and a domestic "cheap food policy." All of these characterize the federal relationship with the agricultural sector, and this relationship leaves farmers feeling very confused. What is the intent of federal farm policies? What does the government and what do other Americans want from them?

Federal farm programs are another mixed blessing in the economic survival of farm families. Although some farmers suggest that they would be better off without any farm programs, most acknowledge that today's market is global and that it is not a free market. The competition is heavily subsidized as part of their countries' national agendas, and our government sees agriculture as much more than simply the production of food and fiber. As a result farm programs seem to take as much as they give. What they take is the farmer's independence as a producer and decision maker, and what they give is low interest loans, price supports, or other types of income. The often competing purposes of farm programs, the sudden shifts in program requirements and support prices, and the farmers' underlying sense that they are caught on a road that is taking them away from what they believe is the essential meaning of farming, all contribute to the mixed feelings about federal farm programs.

The returns to farming are small. In comparison to other businesses, the returns may be minuscule, especially when they are figured on the basis of the hours the farmer puts into the operation. However, unlike other workers, farmers rarely think about their jobs in terms of what they earn on an hourly basis. Perhaps the reason is that to do so would challenge their commitment of time, money, and energy to the business. One farmer who did think about this issue offered the following assessment of his "hourly wages" from farming. "I bought a small farm and milked cows and then sharecropped for about fifteen or sixteen years. I milked cows twenty-two years. Every year I would keep pretty close records. What I found was that it didn't matter if feed prices or milk prices went up or down, I would still make about a dollar an hour. It was just barely enough to keep you from going out of business. For twenty-two years I made it regardless of how high feed costs or milk costs were; whichever way it was, I would still make about a dollar an hour."

Although this may be an extreme case, at a dollar an hour consistently over twenty-two years, you cannot help but wonder, Why go on? Unlike other small-business owners, farmers do not expect to see a positive cash flow each year, nor do they assume that as each year goes by their bottom line will improve. They would like it to improve, but they also recognize the realities of their business. To make it in farming as a business requires confronting and overcoming a set of conditions that cumulatively create an exceptionally risky financial environment. Farmers recognize that they confront a factor of production—weather—that substantially reduces their ability to exercise rational control over their financial success. As a result even the best of farm managers, those exercising the most fiscally conservative decision making, may go bankrupt. It is a situation in which to succeed mentally, you have to be willing to accept that you do not have complete control over your business, and you have to accept the risks that come with change.

The standard strategies used by other businesses to increase their competitiveness are open to farmers, but they typically do not have the same level of benefits for farmers as they do for other businesses. The alternatives are costly in terms of cash, labor, and machinery, and often they simply add new stresses to those that already exist. This farmer explains the delicate balancing act it takes to survive: "You've got to be good, and you've got to be smart, and you've got to be efficient. You probably also have to be diversified. You need a staple like hogs or cattle or something that will give you a little bit of security. Volume is going to help. You spread your fixed costs out and you don't have near as much fixed cost break. But also, when you get bigger, you've got to have bigger machinery and that makes your fixed costs go up."

There are contradictions between how farmers and the public and government officials view the business of farming. The beliefs in self-reliance, independence, and the Protestant work ethic—or more simply, "if you work hard you will succeed"—collide with the reality of dependence on foreign markets, purchased inputs, indebtedness, and the cost-price squeeze. Moreover, most farmers believe that one of the roots of their financial situation—the low profit margins and the cost-price squeeze—is the low cost of food in America. If only, they say, American farmers received the same prices for their food products as their international competitors, then farmers could make a decent living. But they recognize that this is not likely to happen.

Farmers also recognize that there is a certain irony in their economic relationship with the government: the Jeffersonian ideal of the self-reliant and independent American farmer relying on the government for some surety for economic survival. Yet without federal farm programs the loss of farm enterprises over the last thirty or forty years would have been even greater.

The long-held Jeffersonian image of farming as the backbone of the American economy collides with the reality of the increasing economic marginalization of farming. Divergent images of farming as a business appear on national television. Huge machines crossing endless acres of golden wheat and pioneers on the biogenetic frontier, bringing new products engineered in laboratories to America's tables, are counterposed with Ma and Pa Joad eking out a subsistence living on a small parcel of land, poor but proud. These are the essential contradictions in the only "unskilled" job that requires a master's degree in science and a million dollars to perform.

Considering the trade-offs and contradictions of the business aspect of farming would lead many persons to ask, "Why would anyone continue doing this physically exhausting, mentally stressful work for so little return?" It's a question that we asked, too. And the answers, although phrased in different ways, all tended to emphasize the fact that farming is not just a business, it is also a way of life; the one cannot be separated from the other. This fact is the key to understanding farming as a business. "So many of the farm magazines we pick up are written on how to produce more. For a lot of farmers that is the name of the game. But as far as we're concerned, there is no way to a have a family farm strictly by the books or the rules that would apply to any other type of business. Farming is risky, it's a gamble, it's that little spark of unknown. That's probably the reason we farm. But to say we like farming because it's a gamble does sound a little crazy. It's frustrating too. Some days you think, How did we get into this? This is the dumbest thing we've ever done. But then the next day you feel like everything will fall into place and we'll be on the road to success. With so many jobs there's not this kind of job satisfaction. The challenge of farming is important."

Farmers are entrepreneurs, and as entrepreneurs they are risk takers. With risk comes the possibility of failure as well as the possibility of great success. Farm publications frequently carry stories of the farmers who have used new strategies to establish financially successful businesses. Although most farmers recognize that the odds are high, a deep faith in the possibility of success, in the endless possibilities of the business, is at the heart of their harvest of hope.

7 Public Work

PUBLIC WORK

Once upon a time in a land of dreams
chickens laid colored eggs
gathered by a brown-eyed boy
    with an Easter basket,
a lass with platinum braids
picked alfalfa from the fields
    to feed her rabbits
and Belgian mares pulled an Owensboro wagon
    green, with great red wheels
now dusty in the barn.

There was such a thing
as mornings to linger
    by the stove
    in the garden
    on the back porch swing
rides to the hay field
dips at the spring.

It should not matter, darlin'—
farm women serve
    pot roast at noon
    clear iced tea and home made biscuits
    water in the field.
Drive thirty miles
where headlines clip
    and deadlines fly
    every day.

Just don't look back when you pull out the drive
and smoke is curling out your chimneys
   where wood stoves heat silent rooms
and the sun streaming through old glass
   casting patterns on the floor.

Drive, darlin', and remember
farm women do
what needs doing: build a fire
   when you get home
put the soup on
   try
to stay warm.

BASIN SPRING FARM

The work never ends. But now it is more than long hours. It is an eight-hour working day, plus commuting half an hour or more each way, and coming down the drive at the end of the day (or night) to home, family, and farm.

The super-working-woman, mom-wife of urban and suburban exploits pales when compared with a farm wife and mother who also works off the farm. It makes me wonder, Can this truly be done? Can we commute to public work, keep the dust balls at bay and the clothes washed and the garden weeded, keep the livestock fed and the books balanced, and still pick up the kids from band practice on time, put supper on the table, and manage time for our families—let alone for ourselves—before falling into bed at the end of the day to wake up and do it all again?

Can it be done? Should it have to be done? Can we manage it all, keep our health, and still savor the joys that are unique to farm living?

The issue of public work raised a thorny head in 1978, when Jim read the handwriting on the wall. A major recession in the agricultural economy was shaping up, and we were already feeling its grip at Basin Spring. There were other signs, too.

For me, the precipitating factor in taking an off-farm job was that I was tired of watching the roof leak. For six years the turkey roaster had resided in the upstairs, catching the creosote and water that wept periodically down the chimney and sometimes dripped into the strategically placed pan. By the fall of 1978 the drip stains on the ceiling divined that more pans were in order, or we would have to come up with a plan to do something about it. Expenses were mounting. With two children approaching school age and braces and cheerleading and band and everything else that kids today need and want, a decent roof and flashing was a small, if annoying, blot on the financial picture. But it illustrated for me a fundamental principle of farm economics, well known to all other farm women: "If it makes money, fix it; if it doesn't, wait." And wait, and wait. It is the historic adversarial relationship between the house and the barn. Houses don't make money.

So began the quest for the an off-farm job, something to help pay the bills and something I, and we, could live with. First, it was as a 4-H agent with the Cooperative Extension in Breckinridge County. But within the first year I had put twenty-five thousand miles on a new car going to twenty-six club meetings in eight schools throughout the county with meetings at least two nights a week and often on weekends. It was very demanding with small children and the farm. Not the perfect job. So I searched again. A part-time job at a new county newspaper gave me flexible hours, however, seven months later my paychecks started bouncing, and I had to resign. Back to the search.

It took several months, but I found a full-time job as lifestyle editor at a daily newspaper thirty miles from home. The salary was only slightly more than I had been making part-time at the weekly, but the assignment seemed tailor-made. Plus, health insurance was in the offing. I leaped at the opportunity.

For three years the fast-paced world of a daily newspaper and the seasonal life of the farm made strange bedfellows. But I loved my work, the experiences it afforded, and the people I was meeting, and I often wrote about agriculture. I ricocheted between disparate worlds and was often wrenched between time warps. But I managed to juggle my worlds until the spring of 1984, when I was offered the choice of a twenty-dollar-a-week raise at the newspaper or a one-hundred-dollar-a-week raise to work at Fort Knox as a designer of military publications for the Armor School. After eight and a half years, I took another position as graphic designer for the U.S. Army Recruiting Command, and then moved into Army advertising. With this position I traveled frequently around the country, averaging thirty days or more on the road for meetings and photo assignments. I'd leave our nineteenth-century wood-heated farm home at 5:00 A.M. and be on Madison Avenue in New York by 10 A.M. After a while, the contrasts began to overwhelm me. I moved on to my next opportunity.

In all these jobs my life seemed to be on a runaway train. Gone was the seasonal ebb and flow that I had learned to love on the farm. Hurtling through the weeks and weekends, I was determined to hold onto the vestiges of farm living, even though it meant getting up at 6:30 A.M. to weed the garden. A Saturday morning might find me running the washer while I vacuumed, picked vegetables from the garden, and cooked for a hay crew, as my children ran in and out the back door. Our early life at Basin Spring, when I gardened organically, canned, churned butter, and cooked for hired hands, had been compressed to accommodate a more-than-full-time job. I wanted my farm life back—or as much of it as I could retrieve—but I also wanted to continue helping with the cash flow. Is this possible?

It's now been more than sixteen years since I began working off the farm. For over ten years I have commuted from a central time zone to an eastern time zone. I know I have been fortunate to have "good" jobs—and so has Jim, who took a "temporary" off-farm job in 1980 working with the Farmers Home Administration (FmHA) and now works full-time with farm accounts at a local bank. But the price in time and peace of mind has been high. For three years after we had both entered public work, we continued to do all the farm work with the help of a hired hand. "Vacation" leave was scheduled around cutting hay or housing tobacco or whatever the season required. We were always working—on the farm or off the farm—or driving from home to work or driving home to work. This schedule offered precious little time for our family and even less for outside interests and friends.

Over the years we have streamlined the farming operation to accommodate two work schedules. Now we rent our tobacco base out and furnish expenses, and our partner and his family furnish the labor. It is a big job acquiring the tobacco leases to grow a 9-acre crop with a 2.2-acre base of our own. Jim handles that. In addition we rent 160 acres of crop ground for corn, soybeans, and wheat, and we split expenses and profits with another partner who provides the labor and equipment. And we maintain a herd of between twenty-five and fifty head of cattle.

The division of labor on the farm is still basically traditional. Jim and our son, Gerard, do the outside work; I manage the house with biweekly paid help. The help I have buys me the time in the garden, which is smaller than in the past but is a necessity for me, nonetheless. Canning, which I enjoy, has taken a back seat to freezing. I even freeze tomatoes, cored and whole in plastic bags. Jim and I share the cooking,

It is a workable system for now, but with fewer of the pleasures of farm living and all of the responsibilities. There is always mowing to be done—the yard or the pastures—always fence rows to clip, fences to mend, cattle to monitor, and machinery to maintain. All at the end of the "work" day. There is little time for relaxing, and little for socializing. And where, oh where, is the time for jobs that other people ostensibly do when they come in from work, such as patching the porch roof or sealing the flashing around the chimneys or working in the yard? I do not know, but one thing is for sure, what I do not do in the fields I make up for in maintenance. Every Veteran's Day I climb on the roof to prepare the chimneys for the winter months. At other times I have served as "hod carrier" for a brick layer, scurrying up and down two-story ladders for tools and water and sandwiches. I have become adept with flashing and mortar and keep a five-gallon drum of roofing tar indoors so that it will always be at least room temperature if I need to use it. And I keep a jar of wheat paste mixed up in the refrigerator for wallpaper patching.

We all work, and that is good. But sometimes I wonder what kind of farm and family life is possible under these circumstances. Often I am over- whelmed, for seldom, it seems, are we doing enough or doing it right. How do other farm families manage to do this and survive? These questions led me to write about this subject in my work for the newspaper and as a freelance writer. Frequently I have heard my questions echoed in the voices of other farm women.

A friend of mine described her off-farm job as "a force, not a choice." We talk about the loss of time—time to take care of our homes, time to pre- pare "good meals," time to garden, time to can, time to visit with family. We both feel that what we do, we don't do well. And oddly enough, we both miss the sweet interlude of hanging clothes out on the line to dry. We miss our "day off" each week. And we miss the meaning of the seasons and working with our husbands.

Many farm families manage these two worlds, but like us, they have to make accommodations. Some have tailored their farm operations to suit their many jobs, as we have done. What was once a 700-acre grain operation with 30,000 pounds of tobacco and 100 head of beef cattle for one couple we know has been scaled back to 400 acres of hay, 20 acres of corn, 7,000 pounds of tobacco, and a small herd of brood cows. A large garden not only fills their pantry, but also helps to ease their debt load by providing fresh produce for the restaurant her husband manages.

Our biggest constraint is time and energy. Sometimes you don't get all the tomatoes out of the garden when they are ready, or you lose some green beans. Or you don't get the windows washed or the ceiling fan cleaned. And sometimes the house gets a lick and a promise. Hardly a perfect job, but it is better than nothing, and at least I may have time to get to the gar- den today.

Your priorities shift with the changing cash flow. Wants can easily turn into needs when the things you have been putting off for years begin to drift into reach. "Ah yes," you think, "We can finally do that now!" But then the tractor breaks down or the cows tear down a fence or the bottom drops out of the soybean market, and suddenly what you wanted is gone, replaced by what you need.

There are other hazards, besides the obvious changes in work and fam-

ily roles. The old rules of who does what are no longer securely in place. And then, there is never enough time. But on top of this is what hurrying and juggling and working too much does to our health. The stress builds up without any idea of when or if it will end. Eventually, something has to give. In 1992, Jim had two angioplasties within eight months. When the tests came back showing he had normal cholesterol and blood pressure, the doctor asked about stress. Jim said, "Stress? Nothing more than usual. Nothing I haven't always lived with." And sometimes what gives is the sense of an anchor in our lives. The taken-for-granted ways of doing and thinking don't apply anymore, and you can't figure out what's supposed to replace them. How is everything supposed to work?

I often wonder at what point off-farm jobs drain on-farm growth and potential. Two worlds that must be reconciled. And that takes balance and perspective and energy, lots of it. Somehow we must figure out how to juggle all our lives and livelihoods, while keeping our eye on the big picture. And the big picture is the farm.

Always the farm.

BALANCING PUBLIC WORK AND FARM LIFE

They call it public work. The phrase has nothing to do with who employs them. Rather it refers to the fact that they are working out in the public rather than on the farm—a way of life that characterizes the majority of American farm families. Studies indicate that only about one-third of all farm families match our idealized image of the full-time farmer with a farm-helper–housewife and a farm operation that generates enough cash flow to pay debts, meet current operating expenses, and still provide a satisfactory standard of living. And most of these traditional farm families are older couples or those with very large farm operations. Farm families throughout history have been involved in public work, commingling wages and salaries with farm earnings. But in the last twenty-five years it has become more the rule than the exception. This chapter explores what leads some farm families to do public work, how they juggle their farm and off-farm work, and their views of public work and its effects on their lives.

Farm couples do public work for the same reasons that anybody works— they need the money, they find a job that is interesting, they want to do something outside the home. The difference is that many farm couples who work off the farm actually are moonlighting from their regular job— farming—by getting a second and sometimes a third full- or part-time job off the farm. For others, farming is their second job, that is, they see farming as a second career that they are working at part-time right now but plan to move into full-time when they get the farm "set up the way we want it to be," when they reach their forties or fifties.

All farm couples who do public work, like Barb, want to find the perfect job—one that pays well and has benefits and work hours that fit the ebb and flow of farm production. But just as one cannot plan a profit in farming, one cannot plan the perfect job. One job has flexible hours but paychecks that occasionally bounce; another is close but has no benefits and pays little. You learn to make do and accept the trade-offs forced on you by the less-than-perfect off-farm job. Work and careers are not necessarily free, rational choices, but rather events that make sense in a particular time and place. At one point Barb had to decide whether eighty dollars more per week was worth a more rigid work schedule in a different time zone. For some people, living in one time zone and working in another would raise the specter of someday meeting yourself coming and going, but for Barb it is just one of the mixed blessings of being able to find a "good" job in a rural community.

Economics is the principal reason that many farm families do public work. The need to generate additional economic resources to support the family and maintain the farm operation has led many farm husbands or wives, or both, to enter public work. Financial instability—wide fluctuations in income from year to year and episodic or sporadic income during the year—is the term that best describes the economic nature of farming. As a result having a stable source of off-farm income is a strong attraction of public work.

Economic survival, then, has in fact driven many farm families into public work. But this is not a recent phenomenon. The censuses of agriculture since the early 1900s show that the number of farm operators who also have off-farm jobs has steadily increased over the decades. Studies on the employment characteristics of farm households show that between one-third and one-half of all farm families have at least one family member working off the farm. The type of farm operation, the types and kinds of jobs available in the area, and the educational attainment of the spouses influence how likely it is that either spouse works off the farm. If a farm couple looks at a list of their neighbors who have at least one off-farm job, they find that often it is the wife who has the job. Indeed, the sense is that not only is a farm wife's working in town accepted, it is "almost expected for most of them." Off-farm work is one way in which everybody can contribute to the economic survival of the farm. As one farm father explains, "Both have to work to make a go of it." Another farm wife, after explaining that she had never worked off the farm, stated that her son's family life will probably be quite different. "I think most of the farmers' wives do public work. It is almost a necessity today with living expenses much as they are and with what a farm operation makes. In the situation he is in now, if he were to get married, he would never get his farm paid for unless his wife was working too to help make the living expenses. If she worked and they could live off of what she made and he could take what he made off of the small farm and apply it to paying for the farm, they could stay in farming. But it would take both of them working."

Many other farm couples echo this idea. The wife does public work to pay for the family's living expenses and the husband stays on the farm working to make it generate enough income to support itself. As one farm wife jokingly explained: "I went to work for the luxuries in life—food and gas and electricity—just for the luxuries!" In effect, a working wife is the financial edge that permits the family to stay in farming. One farm wife explained why she went back to school and took a public job: "Why? Because I hated not having any money. And we didn't have any. You know, we had enough but, well, we didn't have enough. We had to drop our health insurance one year and we went a whole year without any health insurance because we just couldn't pay it. We could barely pay the light bill. My job does help. It helps tremendously not to have to worry about how to pay the bills each month."

The same need to provide a stable income for the family has led many farm men into public work. The impact of the decision that the husband will work off the farm is different from the effect of the wife's working off the farm, since it is more likely to entail major changes in the nature of the farm operation. But in many communities the husband will be able to find a job that pays a higher salary and has better benefits than will be available to his wife. However, the decision means making trade-offs, such as deciding whether to invest your time and energy into working to build up your own business or as an employee for someone else.

Sometimes it is the husband who may seek public work because in the life cycle of a farm and a family, the highest costs come at the same time. Just as the young couple have begun their farm operation, making significant investments in land, machinery, and livestock, they are also starting their families, needing to meet health, food, clothing, and other living costs, as well as putting some money aside for future educational needs. Although the economic pressures are great at this time, there are also young children at home, and many couples would prefer that the wife stay at home. On the farm, the tradition of the housewife mother is strong. For these families, having a wife do public work is not a desirable option.

This farm wife explains why she chose to work part-time instead of full-time. "We probably did without a lot and cut corners so I would be able to stay home with the children. I don't regret that. It usually takes everything you have and then some to raise kids regardless of what you have. Had I worked full-time when the children were small, I probably wouldn't be much better off than I am now. At least I have the satisfaction of saying that I raised them. I don't look back with regret and say: 'Well, if only I could have been home with them.'"

Finally, sometimes it is the wife who seeks the job off the farm because there remains a belief that farming is *man's* work and therefore it should be the woman who finds work off the farm. When they are farming in a multifamily operation, moreover, it may be a family business, but it is his family business, his career, not hers. In such cases a farm wife may feel a financial responsibility to the family business, but it does not mean that farming is her job to do.

Although many farm couples speak of going into public work in terms of "necessity" or "being forced off the farm," a nearly equal number never intended to farm full-time. They always have planned to do public work for a period of time and then go into full-time farming. In other words, public work has been their career, one that complements and helps them maintain their farming way of life.

Many farm wives had a public job before they married and never intended to leave their jobs. For a town wife, farm work may require skills that she has not had the chance to develop. A public job, on the other hand, offers both a career and a chance to contribute to the household income in ways that best suit her talents and interests. Furthermore, a public job is an opportunity to break the physical isolation of farm life, to visit with other members of the community on a daily basis. For town wives, who have grown up with across-the-fence neighbors, social interaction may be as important as any income that comes from an off-farm job. "I had this job when we got married so I just always kept it. I used to work full-time, but when my son was born I just started working three days, and I've just worked that for twenty-three years. When we first got married we needed the money. Now my husband says it costs more for me work than it does for me not to work. But I like getting away. I like being with people and just having other interests besides the farm."

The desire or intent to hold multiple jobs—public work and farming— is also woven through the comments of many farm husbands. Opportunities for personally and financially satisfying public jobs presented themselves, and it seemed to be possible, in a sense, to have their cake and eat it too. "Right out of high school I got a job in a factory. I worked in the factory about four or five years and continued my education. Then I got a job with a local financial company and then moved to a new company. But all down through the years, ever since I got out of high school, I've always had some type of farming activity on the side. It just seemed to all come together."

Others began their adult life with a clear plan to work full-time off the farm, keep their hands in farming, and eventually "retire" into farming full-time. For these individuals, their primary occupational identification, what defines who they are, is their public job. Although not farming full-time, they have "kept their hands in farming," and so it remains integral to their sense of self. "Public work kept us from going broke on the farm. I've got a good job. I'm the plant superintendent at a factory. I didn't plan to farm full-time until I got ready to retire. My goal has always been when I was fifty to retire and start farming full-time. And I'm probably pretty close to being able to do that."

For some it was clear that the opportunity to farm would not be there unless they could find a full-time job that would help them buy their way into farming. Farming is expensive—the land, the equipment, the livestock, and then the wait for the return from your time and investments—and getting the funds to start farming often means finding a public job that can provide a steady income. In some cases, the full-time job became permanent, but in others the job lasted just long enough to gather the funds to break into farming. "There's no way I would have started farming without my public work. I had too much income for FmHA [Farmer's Home Administration] and I didn't have enough down payment to get a bank loan. I had enough money for a down payment on the farm, but I couldn't afford the equipment and everything else I needed. So I stayed in public work until we got to the point where we were in a financial position that we didn't have to worry. We got things here on the farm fixed up to suit us. We bought this farm and we paid for it."

Clearly, the reasons for doing public work are varied, but several common themes run through all of these comments on the decision to work off the farm. First, the desire to remain in farming, to "keep their hands in" even while continuing a full-time career in public work is strong. Second, there is uncertainty, especially among those who have worked full-time jobs, as to how having an off-farm job affects their position as *farmers*. And finally, they realize that off-farm income gives them some financial security, but not without some costs. It was that fact that haunted Barb and Jim at Basin Spring Farm. Barb has raised this issue many times in this way: "Though farm families are shouldering outside jobs in increasing numbers, the overriding question is, Can we live well? Can we manage it all, keep our health, and savor the joys that are particular to farm living?"

Despite the economic pressures, about one-third of all farm families do not have off-farm jobs. Do they remain on the farm because they do not need the extra income? Everybody could use more money, of course. But for many farm families the question is, On balance, is what we gain in additional financial security worth the costs? How each farm family answers this question illuminates some of the potential consequences of working off the farm.

Although many farm couples do not have public jobs, most of them have thought about getting one at one time or another. But the loss of an unpaid family worker on the farm is a powerful counterargument to public work. If the farm needs all the hands to keep it running, the cost of replacing a spouse's labor may be more than the wages that can be earned off the farm, especially if the one to take the off-farm job is the wife. The earnings gap between men and women in the off-farm marketplace means that for some farm families it is not profitable for the wife to take an off-farm job. A young farmer explained how he and his wife weighed this decision: "Things are getting tight. The wells are dry and worse is yet to come. I just mentioned that maybe she could think about getting a job. If we could live on what she made on a job off the farm, then that would allow that much money to go back into the farm. At the same time, I know that she's a very important part in the whole operation. She goes down and feeds twenty calves, takes care of graining and watering. Plus everything else she does. If she got a job off the farm, we'd have to hire somebody full-time. I've told her that she is worth more here than she would be with a job in town. It might depend on how good a job she could get off the farm, I guess. If she got a good enough job off the farm, that might be the way to go. But see, the pay scale's not very good in a small town."

What are the costs of staying out of the labor force or taking a particular job versus the potential returns? Economists call these "opportunity costs." Wage rates in rural areas have always lagged behind those in cities, and when you add in the earnings gap between men and women, you have a rural labor market that offers less to all workers than they could find in a city, and even less than that to women. This is one reason why families who farm near large urban areas are more likely to have off-farm jobs than those who farm in more remote areas. For many farm families, the labor demands of the farm—whether other labor is available for hire and the cost of that labor—must be balanced against the potential earnings from an off-farm job. In many rural communities either it simply is not possible to replace a spouse's labor on the farm, or the costs of doing so are greater than the wages that a spouse can bring into the household from public work.

Also influencing this decision is the nature of the farm operation itself. There are certain types of farm commodities that are less compatible with off-farm work. For example, ideally, dairy cows should be milked three times a day to maximize milk production; at a minimum they must be milked twice a day. The labor demands of dairying make it difficult to consider public work. "Well you can't work off farm if you're milking. If you run eighty milk cows, to get out and work in town, you have to hire somebody else to help with the milking. To get anybody that you would turn loose with a bunch of milk cows, you'd have to pay them a lot, probably more than you can make."

This is not to say that farm couples have not somehow managed to dairy and do public work. They do, and the kind of juggling of farm work and off-farm work that it takes to do so successfully exemplifies the challenge of blending public work and farming. As at Basin Spring Farm, it involves setting priorities. What do I spend my time at home doing? What do I stop doing? How do I deal with the guilt of stopping or letting slide the jobs that I once believed needed to be done but now do not have the time to do, or to do in my usual way? The extent to which farm couples are willing to juggle multiple jobs to maintain that delicate balance between doing well on the farm and doing well in the off-farm job illustrates the strength of their commitment to farming as a business and a way of life. Several couples describe a typical day:

A dairy farm wife: "I would get up at 2:30 in the morning. Milk the cows, come into the house to clean up, and then drive eighty miles to Louisville to work. Then when we came home, we'd go back to the field and milk the cows. For twenty-six years we only worked on the farm between 2:30 and 4:00 in the morning and after 6:00 at night."

A cash grains farm family: "We both worked public work for all our lives. Our farm is all grain and tobacco with a few beef cattle. He would get up and check the animals and then go on to work. I would get the kids and myself ready and take them to school and then go on to work. I try to call him during the day so he can tell me where he's going to be after work. So he comes in and changes his clothes and he goes and combines until about nine or ten at night, and if I didn't save any dinner for him, I cook his dinner. He always works the farm on Saturday and Sunday too. He gets out early Saturday morning and he stays gone. But on Sunday he always stops long enough so we can sit down to eat at the same time, and the rest of the time it's hit and miss."

These descriptions of typical workdays combining farming and public work were echoed by many other couples. It clearly is not easy to balance two full-time jobs, especially when the labor demands of one are not set by the time clock but by the seasons and happenstance. The plans to top tobacco are interrupted by the cows that broke through the fence, and the hay field that must be cut must wait till the fields dry from the afternoon rains. Such juggling creates stress that has physical consequences. The dairy farm wife who described their multiple careers is a widow. Her husband had a heart attack within a year of retiring from his factory job. She continues to farm but admits that it does not have the same meaning for her as it did when she and her husband were sharing the work. For a farm wife it often amounts to three full-time jobs, because besides her public work and her chores on the farm, there is still the housework to do. This simple recitation of when all her work was done was repeated by nearly all the farm women with off-farm jobs: "After work I would come home and feed the livestock while my husband was out in the fields with the boys. I'd cook and they'd come in around ten and we'd eat dinner. I'd do the dishes and then start the wash or the housecleaning. I'd get to bed around midnight."

Some couples manage to keep all their balls in the air, juggling their multiple work demands fast and furiously while ricocheting from farm to public work and back to the farm. But others, like those at Basin Spring Farm, realize that the frenetic pace cannot continue forever; something must give. In some cases it is the off-farm job, and in other cases it is the farm as another farmer who used to do public work explains: "I think the last year I worked at my public job we still had thirty sows and thirty cows, we were raising 3 or 4 acres of tobacco, and I had 350 acres of row crops. I was still working forty hours a week. But at that point I had to make a decision to either go into farming full-time or back off. I couldn't take it any more, I just never did have a minute."

Another farmer explains why he got out of dairying and into beef cattle: "I didn't think I could dairy when I came back to the farm because dairying is too labor intensive. I didn't feel that I could do a good job as a dairyman and work off the farm. But you could manage beef cattle and do public work."

Quitting the public job or scaling back the size or the nature of the farm operation are two ways of regaining control over their lives. For others, it comes with the "luck" of finding the perfect hired hand, one you can trust, who is capable of handling the work, the equipment, and the decisions. "After I hired my full-time manager we expanded. We got up to over one thousand acres of grain and also started producing alfalfa hay to sell. I thought that I had to because the farm had to generate enough to pay him full-time wages. We couldn't run the operation without him. He is sort of a one-in-a-million kind of person that just does everything, makes all of the decisions. It is not unusual for me to go a month and not talk to him. It is sort of a unique situation. I certainly couldn't operate at the level that I operate without him as an individual. He is a pretty important piece of my farming operation. He maintains the farm and makes the majority of the decisions. We converse periodically. I make the marketing decisions, primarily. With just a regular hired hand you have to be there to tell him how to do it. Well, I can't be here. My job requires me to be gone from home a lot. No, we wouldn't be farming if it wasn't for a full-time manager."

For farm couples doing public work, time becomes a precious commodity. There is barely enough of it to do a good job at both jobs, and once both jobs are done there is hardly any time left for family and almost never any left for friends. One farm wife commented that for several years her husband rarely saw their boys awake. He would be off to his public job before they awoke. He would be out in the fields when they came home from school and not get back to the house until after the boys had gone to bed. With some sadness she noted that he had missed much of the joy of watching his boys grow up.

For farm families doing public work, vacations become a memory. Vacations are simply a chunk of time to catch up on farm work that has had to slide because it required too large a time commitment. This couple describes a "typical" vacation that sounds very familiar to many other farm couples doing public work: "We spent our first vacation after we bought our first fifty acres fencing it."

Trying to manage the time to do too many jobs creates enormous stress. It is stress that piles up on top of the stresses that are already associated with trying to make farming profitable, to work comfortably with relatives, and to raise children successfully. It is stress that often is rooted in the gnawing sense that you are doing too much and not much of it very well. And it is a mental stress that comes with always being tired but rarely having the satisfaction of a job well done. "It was pretty rough. Of course I was young and I'd get tired and with a good night's rest, I was ready to go the next day. But it got me mentally too, because a lot of days when I knew that I needed to be doing something on the farm, there I was working at the factory. The mental thing was probably harder on me than the physical, although the physical was bad enough, because I didn't have a minute. Whether it rained or shined, I had work that I had to do, whether it would be in the hog barn or working on a piece of equipment or out in the fields."

Farm couples see the toll of juggling multiple jobs in their own lives and in the lives of their friends. The small things that used to bring plea-sure—like gardening or hanging the wash out to dry or quilting—are pushed aside by the lack of time or simply the lack of energy. Sometimes it is hard to keep an "eye on the ball," to remember, in the rush of things, what it is all for—keeping the farm and the family together. The reasons for putting up with this hectic pace are clear when couples describe the advantages that public work brings to offset some of its costs.

If economic necessity leads many farm couples into public work, then it is the answer to this need. As many couples have already noted, public work can provide the money to buy their own farm and give them an opportu-nity to own their dreams. But the money from public work is used in many ways. It often means the difference between scraping by and having a modest standard of living. "If it wasn't for me working off the farm, we wouldn't have a lot of the fringe benefits or other things that we can af-ford that a lot of people don't have. For example, it would have been aw-fully hard getting the kids through school, because we had three in col-lege at once. If you have a double income, then you can have these things that are hard to come by on the farm, especially if you are both paying for the land and living off it."

But keeping off-farm income for family and household expenses as opposed to the farm is not easy. The age-old "adversarial relationship be-tween the house and the barn" continues today. We finally have the money to fix the roof on the house. But the tractor broke down. There goes the money for the roof. Some couples strive to keep the moneys separate, to use public wages for the family and the household and make the farm support itself. But it is not always possible. "The first year he worked at the warehouse we put all his salary back into the farm to help pay for this and that. But then we decided we weren't going to do this any more. If he was going to put in that many hours off the farm, he was going to put it into something that we could have. So we've got a separate farm account and a personal account. But we have used part of the personal account for the farm because it is hard to keep things separate. For example, we've got some insurance plans we pay for from the personal account that is actu-ally covering part of the farm too."

Not only does public work mean extra money, it means extra money on a regular basis. By and large, farm income is episodic: it comes when you sell your crop or take your livestock to market. The problem is that utilities, food, gas for the car, and most other expenses come every couple of weeks. The result often is a cash flow shortage. Public work can solve this problem. "The pay is not so good. But I do know that I am going to get a certain amount of pay at a certain time. On the farm you may not. You might put it all out there and not get anything back."

But the advantages of public work go beyond regular wages. For many farm families, having a public job means having family health insurance at a price they can afford. The cost of a privately paid family health insurance plan can be three to four times that of similar coverage obtained through a group plan. Although farm couples often can buy group insurance through a farm cooperative, prices are still higher than for other group plans, because of the higher accident rates and the higher average age of group members. Health insurance through an off-farm job can save a farm family as much as three hundred to six hundred dollars a month. "I've got group medical and dental insurance for myself and the family. That is worth a lot because you are looking at about $160 a month that they are paying for me and my family. We have life insurance too. That translates economically. Just put that on top of what is your actual salary too. I'd hate to lose my job as much for that as for anything else."

Another financial advantage to public work is the opportunity to set up a viable retirement program. For most farm families, their land, buildings, equipment, and livestock represent their retirement nest egg. Few have IRAs or private retirement accounts. Their retirement security depends upon farm values, the nature of the farm market, and the inheritance arrangement with their heirs. The security of their retirement, in other words, is highly dependent upon conditions outside their control. A public job can provide greater retirement security, offering a financial incentive beyond the salary. "I had a fantastic job offered to me, but I had to move to D.C. We really contemplated that. I could make in one year what it would take me three or four years to make here, and it had a pretty good retirement plan. I could have spent ten years in that job and come out with a good retirement. That is what we are really looking at. Here now on the farm, our retirement has gone because of the depreciation in the farm value. Our retirement security has vanished. That's what bothers me more than anything now."

Money for personal expenses, money for your children's education, money for some of the good things in life, money on a regular basis, health insurance, a retirement plan, are all economic benefits of public work. But always detracting from the obvious advantages of these aspects of public work is the underlying stress of trying to get all this by juggling too many demands. "The outside income was necessary. We built a house, and the outside income paid for the house. It was paying for our living expenses and insurance. I never had to worry about us having food or a car or something like that. The outside income took care of the personal life. The stress was coming from trying to hold the farming operation together and keep from going bankrupt. The farm had to carry the farm, and a lot of times it barely could do this."

Blending farm and public work gives farm couples a unique perspective from which to see the contradictions or mixed blessings inherent in these two types of careers. Indeed, such comparisons highlight some of the things that both attract farm couples to public work and repel them. Although extra income and the other financial benefits of public work are important, for some individuals these economic blessings are supplemented by noneconomic benefits. But, like all good things, they also come with some costs.

Public work has benefits besides wages, health insurance, and other economic perquisites. Rural areas typically have a limited number of major employers, but there actually is some variety in the types of jobs that are available. Farm couples work as teachers, bankers, factory foremen, accountants, loan officers, sales clerks, beauticians, warehouse workers, insurance salespersons, truck drivers, heavy equipment operators, and factory workers. Each of these jobs can offer unique challenges, ones that are different from what farmers confront every day on the farm. As a result many indicate that what attracts them to public work, or what has kept them at it, is the opportunity to do another type of challenging job. "I like what I do. I don't know that I'd be satisfied being a full-time farmer. I deal with four hundred to five hundred people every day as plant superintendent. It's just not the same. I don't know that I'd be happy being a full-time farmer."

Another attractive feature of public jobs is that there is a clear beginning and end to the workday. Unlike farming, where one often has to keep on until the job is done, whether it is getting the crop in before the rains come or staying with a cow having trouble calving, most public work starts and stops at specific times. Such a schedule contributes a regularity to one's life, and there is comfort in routine. Family life can develop customary traditions—eating dinner together, visiting friends or family on

Sundays—that often cannot exist when your work life is regulated by the seasons and the weather and the rhythms of nature. A routine can be a luxury when you usually do not have it. "Well my Mom and Dad got up and went to work at 7:00 and 7:30 and came home at 4:30 and 5:00. Their day was done as far as their job. Then we would sit down and eat a meal together. When I moved to the farm I couldn't get over how different this was. He'd get up and go milk at 6. And unless I went to the field and took lunch or dragged the kids back to see what I needed to help do, I did not see him. The kids would be in bed asleep when he left, and they'd be back in bed asleep when he got home. They'd go for days and never see him. So I guess the thing that bothered me was there was no routine."

When work has a set beginning and an end, at a place different from where you live, you can leave your job problems behind as you walk out the door. This is a luxury that farming does not offer. When the drought comes, even when you are sitting at your kitchen table eating dinner, you can see your crops dying and your profits for the year withering. You can never get away from your business problems. With public work, problems stay at the office or the plant, or at the very least, you do not have to look at them twenty-four hours a day. "That's the only thing I like about working in town. When you walked out the door you were done. You didn't have to think about it all of the time."

One reason why you can leave your work problems behind with a public job is that as an employee you typically do not have anything personally invested in the business. When you are an employee, you do not have to buy the machine you operate or pay for the building you work in; someone else pays those costs. If there are problems, they are someone else's problems, not yours. "When you do go into a factory, you go in at a certain time and haven't got anything but your time to put in. You don't have any expense other than getting there."

But, as one farmer noted: "It is not what it is cut out to be, public work." The old adage "There's no such thing as a free lunch" is true. Those characteristics of public work that attract farm couples also are seen as its drawbacks when compared to the aspects of farming that most attract them. In other words, the catalog of the things that they do not enjoy about public work says as much about the nature of farming as it does the nature of their off-farm job.

The first problem with public work is that it is public; it is away from the farm and the farm home. For persons who have been raised in a farm household or who entered into marriage expecting to live and work in the same physical space, having to spend a considerable portion of their day away from the farm and the family makes a mockery of the very things that attracted them to farming in the beginning. It is as if an implicit contract with life and your family has been broken, as described by this farm wife: "I worked at this company after we first married. I hated it. I left at five in the morning and got home at five in the afternoon. In the wintertime I wasn't home during the daylight. It was dark when I left and dark when I came home. I just hated that. I would rather do without things than have to be away from my family for that length of time."

As Barb came to ask, Is this why you went into farming? What about the opportunity to work side by side with your spouse and your children? What about Dad being able to come in for lunch with you and the children? Public work is in town, off the farm, away from those things that make farming and farm life meaningful. So the extra income, the access to health insurance or retirement benefits comes at the cost of having to spend at least eight hours a day off the farm and then even more hours every night and weekend away from the family catching up on farm work.

Moreover, although some people are lucky enough to have a public job that is challenging and personally fulfilling, not everyone is so fortunate. A factory job, a job that involves continuous repetitions of the same activities, is often the only job available in a rural community. When compared to the constantly changing nature of work on a farm, many farm people feel that a lot of public jobs are simply boring. "I worked in a factory for seven years, and I always wondered on Monday if I'd still be there on Friday, because I hated it that much. I didn't like doing the same thing over and over and over. I knew three days after I went to work there I'd made a mistake, that I couldn't continue that kind of life—going to bed that tired of a night and thinking you've got to do the same thing over again the next day."

A factor that contributes to the feeling that public jobs are numbing and routine is the timing and sequencing of much of the work. Work off the farm is compartmentalized. There is a clear beginning and ending to the work day, and in many jobs there are set times to take breaks. When compared to the timing of farm work, in which the day ends when the work is finished, public work is rigidly scheduled, and each work setting creates expectations about what a normal work day should be like. Consider: a farmer anticipates as part of raising livestock that he may spend all night with a cow having trouble calving. But secretaries do not expect their supervisors to call with a job request at eleven o'clock at night, because their job ends when they go home. What some see as the advantage of the routine scheduling of work, others see as a senseless absurdity. "That was something I never could get used to in the factory. When that whistle blew, you sat down and took a break. It didn't make any difference what you were doing. That was the law. You took a break. When the whistle blows, you go to lunch. If it would have taken thirty seconds longer to finish up what you were doing, and it could have saved thirty minutes later, you still had to quit when the whistle blew."

Another major difference between public work and farming is what it means to be an employee. When you work for someone else, you may not have the investment in the job that you have when you farm. But you also do not have the satisfaction that comes from doing creative work; you are expected to be an obedient and reliable employee. Most farmers believe that the balance sheet at the end of the year reflects how well they have *personally* done, *both* as manager and as worker. When you see your corn standing over your head, full of perfect ears, you know that it is the result of your hard work, your decision making, and you can take pride in what you have accomplished. But the sense of pride in a job well done can be missing in many public jobs, and in others the source of pride and accomplishment is fundamentally different. This difference is often a source of dissatisfaction with public work. "I worked in a factory for twelve years, and one of the reasons I quit was I didn't like doing the work and somebody else taking the credit for it, moving on up the ladder, and leaving me to set there. Also, it bothered me that I was not able to make my own decisions, being able to do things the way I saw they ought to be done. And when I did do something on my own, I knew who was going to get credit for it. It wasn't me."

In most public jobs the labor is separate from the authority to make decisions, and the responsibility for the quality of the job is vastly different from that in most farm enterprises. Many farmers believe that how well you do your public work has little to do with how much you earn. Ironically, although the same can be said for farming, many feel the situation is somehow different. "With so many jobs there's not much job satisfaction. You just do the forty-hour week, and if you do the very best that anybody could do, you get the same paycheck as the guy who does just about as bad as anybody could do."

Interestingly, there is another aspect of public work that many farm couples find disturbing—the lack of job security. An advantage of off-farm work is that you have nothing invested in it except your own time, but you also do not have the security that comes from ownership. As one farmer explained, "If you have a public job, you have a risk of them coming and telling you that you are going to be laid off next week or next month or whatever, or their company is selling out."

Now the question of job security may seem to be an odd concern for persons whose livelihood depends upon a whole host of factors—weather, insects, disease, market prices, the exchange value of the dollar—that are beyond their control. It may seem an odd criticism coming from people in a business where, quite often, even the hardest working, smartest managers can go bankrupt. But the criticism goes to the heart of the meaning of farming and farm life to many farm couples. It captures the essential difference between public work for someone else and farming for yourself and your family. In farming, you have the land as your security. "Well, I guess really the backbone of it is that there is not a lot of job security in public work. A person has got to have some base, something they can count on, some security. A piece of property is it. If you've got it, then there is a little more security in your life."

But perhaps the characteristic of public work that is most disturbing to many is one that weaves through all of these comments as a silent thread of complaint. When you do public work, you are not your own boss, you are no longer an entrepreneur. Other people are making the decisions. Other people decide who is to do what, when, and how. Other people decide when you start to work and when you stop. Other people decide what risks to take; other people control the dynamics of the business and the work situation. Put simply, it is not like being a farmer. And for many, the money and the benefits come at the cost of working for and under someone else. Again, there is some irony in farmers, who increasingly find their own decision making circumscribed by government regulations, criticizing public work for the lack of opportunities for independent decision making. In farming you may be your own boss, but not necessarily to the extent that you have full and absolute control over your time, your work, or your business. Yet for many this is an important difference, as clearly stated by this dairy farmer who had started out working as the hired manager for a very large dairy operation: "When I worked at Dairy Corporation, that was a good job. It paid fifteen or sixteen thousand dollars by the time you took the benefits and everything. There was security with that job, and I could've stayed there and eventually made pretty good money. We even had a real nice house. But I just wanted to be my own boss. It's not that I can't get along with people; I just didn't like working for someone else. It just didn't appeal to me. I think that's the main thing about farming. You give up a lot, but you don't have to listen to somebody over top of you telling you that you're doing something wrong, or just not the way they want you to do it."

Public work versus farming. Sometimes listening to the farmers' views of the pros and cons of each, you cannot help wondering whether they are describing the same thing. The characteristics of public work that attract are also those that repel; the characteristics of farming that attract are also those that create a sense of insecurity. To get the positives you have to be willing to tolerate the negatives; it is one of the mixed blessings of blending farming and public work.

Another way that farmers think about the difference between public work and farming is in terms of a career. A career, according to the dictionary, is a chosen pursuit or lifework. Farmers who do public work often struggle with the question, What is my career? Am I a farmer, or am I defined by my public job? The answer is not simple, because although one may provide the bulk of your family's income, the other may be the work that is most meaningful and satisfying. The answer gets to the heart of an individual's self-identity, the sense of who one is. It also gets to the heart of what it means to be a farmer, as this factory manager-farmer explains. "I worked for the factory for sixteen years. So I haven't been a farmer. I have been linked to the farm ever since we've been married, which is twenty-seven years. From the very start I would say I realized that my factory job came first. My job had to be done because that was my income. I knew that my income was not coming from the farm. But my desires, I guess, were on the farm."

The reality is that many public jobs require this kind of commitment and loyalty. If you are a wholesale representative, you are expected to be on the road, servicing your clients. If you are in a factory and they need to increase production with overtime, you are expected to work overtime. Employers expect loyalty from their employees, expect their employees to place the job before home and family. But the farm demands the same and more. Work is not done to a clock; the job is over when the work is done, regardless of any other commitments. The farm takes, but it does not always give back. "If there was a direction, I guess that has more influence, farming versus my manufacturing career, farming is more of a hobby. Yet it is not a hobby either, because it is too much of an investment to call it a hobby. I don't depend on farming to make a living and support my family. If I did, I'd probably have to think about it a little bit differently. I'd be more nervous."

The reference to farming as a "hobby" is interesting. Most people probably would not define as a hobby an activity that requires the investment of hundreds of thousands of dollars to begin and thousands more each year to continue. But perhaps this is the only way to deal with the contradictions involved in trying to decide where your loyalties lie, especially when you realize that you have a potentially financially devastating hobby. What is being suggested is that some people do public work, and farming happens to be something they do primarily for pleasure, not as a business. "The public life has a tendency to wear you down. You want to do the right things in all of the decisions you make. Maybe you worry a lot about making the right decision. Maybe you are caught in a position where there is no right decision. There is a lot of stress there. The farming, in a way, was therapy to help you keep your sanity in a public job."

The dictionary offers another definition for career: a person's progress in his occupation; and here is another point of comparison between farming and public work. In a public job there is an expectation that over time, if you work hard and do a good job, you will have opportunities to move up a career ladder, to earn more each year, to have longer vacations, or to receive other benefits. In other words, in a public job there is advancement and a clear progression toward retirement, with the realistic prospect of being able to retire comfortably. Farming offers no such promise. Ask almost anyone in a farming community; there are not many "retired farmers" around. This farm couple are in their midsixties, and there was a sense of expectations lost in the comparison of their future with that of friends who had left the farm for public work. "We had friends our age that started out working public work when we started out farming. They said: 'Come on, go to the city with us. We will show you a better life.' There is no better life today than working on the farm. They are retired now. They have a life of leisure. We see no future for retirement. We say that we will work until we become disabled and take it from there."

On the standard economic measures of what makes a career desirable, farming does not stack up in comparison to most others. Farmers recognize that fact, and the differences are magnified when they compare their lives with those of friends who went into other careers. But this realization is counterbalanced by a belief that farming as a career offers other advantages that are not present in nonfarm jobs: for example, the opportunity to be your own boss, to build equity in something that can be passed on to your children, or the satisfaction that comes with creating life in the fields and barns. Do these career advantages of farming offset the advantages of a career in public work? The answer is unique to each individual. However, among the middle-aged farmers who have reached the point in their work lives where their expectations and dreams have collided with the harsh reality of the farm business, the evaluation has sometimes found farming wanting. This farmer's analysis was echoed by many others: "I started farming in 1957, and I've done more work every year for less money; every year it seems to get worse. But I would never have been satisfied doing anything else. Sometimes, though, I think if I had it to do over, I'm not sure I would farm. We figured that things would get better. We had some hopes and dreams like all ordinary people. We hoped that by this time in our life we would have a new comfortable home and a few things that other people have. But every year it seems that we won't be able to accomplish these things. There is something wrong when a farmer can work fourteen or fifteen hours a day and can't make a decent living on a farm."

Many other workers in our society would agree. There is something wrong when any worker can work fourteen hours a day at any job and not make a decent living. Many minimum-wage workers are in this situation. So the comparison of farming and public work as careers is difficult. The standards or measures may not always be the same, making it impossible to truly balance what each career has to offer. "If you work in public work, you go in and you have a bad day, you say, oh well. You know, what the heck. I still got my base pay today. I'll do better tomorrow. The farmer don't have that luxury. He's the only guy I know of that keeps on taking a beating and always looks forward to the next year. Whereas someone else would quit and find themselves another job."

Public work—it is a strategy full of contrasts. The farm couples we interviewed could easily catalog the advantages of public work. Off-farm jobs often are an economic lifeline for farm families, providing the edge that they need to keep farm and family together. Public work also gives them access to other benefits they might not be able to afford on their farm income alone, things such as health insurance, retirement plans, and money for their children's college education. And for some, public work provides an opportunity to pursue a job they enjoy, a job that is intrinsically satisfying.

But these advantages must be compared to perceived costs. There is enormous stress from trying to juggle too many jobs with too little time. The small things that have made farm life pleasurable—walking the fields with your children, working beside your spouse—have to be set aside in the pressure simply to get done what must be done. Furthermore, the very nature of public work—the structured work day, the routine work, the fact that responsibility for work conditions and outcomes rests with someone else—make it less desirable as a career than farming for some. The challenges of public work are different and for many, not nearly as satisfying as those from farming.

Public work, then, is a double-edged sword for many farm families. It requires commitment and the concentration of time and energy on maintaining a delicate balance between the demands of the farm family and the demands of the public job. But most of all, as Barb says, it requires always keeping your eye on the big picture. The farm.

8 Farm Families, Neighbors, and Their Communities

BASIN SPRING FARM

The landscape is changing radically in Breckinridge County. The only lights that Jim could see at night when he was growing up were stars in the sky, except for the lights at the neighbor's down the road.

Now the hillsides are dotted with security lights after dark. Jim and his brother recently counted twenty-three lights on Sinking Creek hill alone. Within the last year five new building sites have been bulldozed on prime farmland along U.S. 60 from Sinking Creek to Tiptop on 31W. Twelve new houses have gone up and three trailers have been installed in a field that was rich with corn last season. I never imagined I would have to watch pasture staked and tracked with gravel roads for prospective home builders in this part of Kentucky, as I witnessed in my last years in the Brandywine Valley. But I was wrong.

It is here. The most recent alteration of the rural landscape occurred in the late summer of 1991. For about two years Jim had been interested in some land that lay west of Sinking Creek. We began to entertain the idea of buying it ourselves during the winter of '91. But there were several factors holding us back. It was bad timing, with illness in the family and the farm's finances being somewhat shaky. We could not speculate on buying land unless we could pay it off by farming it, and that was too uncertain at the time. If we had to sell later, we worried whether there would be a buyer willing to keep it in agricultural production. So much good farmland was being sold for subdivisions and shopping centers, it would hurt to see this land go the same way.

So we sat tight and hoped. But in the summer of '91 a realtor bought the rolling valley, and by fall it was crisscrossed with gravel roads. It is unrealistic to think that you can buy up all the prime agricultural land to keep it from being developed, but Jim and I felt sick with what we saw happening. The final irony occurred in the winter of '92. Jim was asked to draw up a loan for the first house to be built across Sinking Creek from Basin Spring Farm. And this is only the beginning. Beside being developed, the countryside is being logged. Or is it being logged and then developed? The result is the same: prime farmland lost to economic development. The question we keep asking ourselves is this: Is economic development a boon or a bane for rural areas?

About ten years ago the county entertained the prospect of having Ashland Oil build a refinery in Holt's Bottom. The topic generated quite a bit of talk. The pros and cons were debated endlessly, and public opinion ran high on both sides of the issue. The pros were plain. I do not know of anyone who would not appreciate a well-paying job close to home with benefits, such as health insurance, paid vacation, sick leave, and life and even disability insurance. A company like Ashland Oil might provide these kinds of opportunities.

On the other hand, building and operating an oil refinery would have a tremendous impact on the county and the surrounding areas. They said two thousand workers would be needed during the construction phase of three years—two thousand people who would need a place to live, plus everything else that goes along with adding that number of people to your community for a couple of years. And then many of them would move on. Breckinridge County would not be the same when two thousand more people and their families moved in, and then left.

But we, my friends and neighbors, didn't have to make the decision. Ashland backed out, because another site was deemed more desirable. Yet the entire episode had graphically illustrated the double-edged sword that confronts many rural communities today. People want to move to our rural communities to have a piece of the good life—room for their children to grow up in and all the other myths that surround rural living. The result is bedroom communities, where commuters live but do not work. But the qualities that make a place so desirable to live make it even more difficult for many of us to do the jobs that are an integral part of what makes rural, rural. Farmers can't afford to compete with developers for good land—land that is good for farming and for subdivisions. So Jim and I watch the new houses go up across Sinking Creek and wonder, Can we continue to farm with urban neighbors? What is going to happen to the quiet rural setting that we and our friends and neighbors have enjoyed?

Another change in the rural economy is the influx of franchised businesses. Too bad for family-owned Gladson's in Hardinsburg or Tobin's Dry Goods. Where can you find sheer cotton diaper shirts with scalloped hems any more? Or have your baby gift wrapped while you browse through work boots and overalls and spackleware and pots and pans? Wal-Mart brought some of these things and more. Jobs for many men and women who go home to house and strip tobacco. Don't get me wrong; I love Wal-Mart. It has everything you need and more. But somehow darting through the aisles of a well-lit, computerized store, just like every other store of its kind around the country, is not the same. It is not the same as browsing through an aisle where the creak of hardwood floors was punctuated by the ring-ring of an antique cash register, where you knew your parents and grandparents had also shopped and visited with their neighbors. It is commerce of a particular kind that we have lost, commerce within community.

We have lost our neighbors in the old-fashioned sense, too. The daily sense. At Basin Spring Farm we know—and our neighbors know—that we could go to one another at any time for anything, and sometimes we do. But more often we see one another in passing—on the road, occasionally at the grocery, or at weddings and funerals, where it is thought, if not spoken, "We have got to stop meeting like this!" We are too occupied with jobs and commuting and getting caught up with chores when we get home and staying abreast of our immediate families to attend to our family of neighbors in the way that was customary when "neighboring" flowed through the course of the day. Then, you might see your neighbor cutting a field of hay while you were mending a fence, and more than likely you would stop to talk and catch up with news of your families. We paid attention to one another. Our attention now is fragmented, because there is only so much energy to go around. You try to nurture what little you have, and community is lost.

Two years ago new neighbors moved in across the hill, and I took over a couple of loaves of warm bread to welcome them to the neighborhood. We have become friends—she and I lost a parent at the same time—but we see each other only occasionally. I send her warm thoughts as I drive by every morning on my twenty-five-mile drive to work, when the sky behind her hill is streaked with light. I am glad she is there. "Perhaps I'll call her tonight and we can get together Saturday morning," I think to myself. But too often Saturday morning turns into dental appointments and grocery shopping and trips to the vet, or one of the kids is home. . . . We are neighbors who care about one another, but this is not neighboring in the community tradition.

We look around Breckinridge County and see all the changes: The new country homes where crops once grew. The stores once owned and run by neighbors, now closed. The strip malls of convenience shops and fast-food restaurants that don't sell country eggs or the handmade, custom sandwiches sold at the gas station-grocery-sandwich shop that once stood at the intersection of two country roads. We see our neighbors' sons and daughters moving away, replaced by those who look at farming as something quaint. And we can't help wondering, Who will farm the land tomorrow? And will there be any land left to farm?

CHANGING FARM COMMUNITIES

Farm families live and labor in a social world that is changing as much as their business. Once isolated by geography and physical distance from the economic and social changes occurring in urban areas, most rural communities now are integrally tied to events in urban America and around the world. The expanding system of inter- and intrastate highways is a physical reminder of the interdependence of rural and urban communities. The highways and roads also link the social and economic lives of farm families with their surrounding and more distant communities. This chapter explores the social world of farm families, their neighbors, and their communities.

There are more than 350 agriculturally dependent counties in America. In these areas a substantial proportion of economic life is centered on providing services and goods to local farmers, purchasing and/ or processing agricultural goods, or working on farms or in agriculture-related businesses. Farming or other types of farm-related enterprises define the economic base of the communities, and the fortunes of the community businesses and citizens are tied to the fortunes of the nearby farm families. The communities in which we interviewed are examples of farm-dependent communities.

Grain County is one of the top grain-producing counties in the state. More than 70 percent of the land in the county is in corn, soybeans, and wheat production, and farms are considerably larger than the state average. Grain County has one of the best grain markets in the nation, and the majority of the corn and soybeans are sold locally, typically for prices higher than the national average. Although the industrial base of Grain County has been steadily increasing since 1950, a significant proportion of all county income comes from agricultural production or agricultural services. An interstate highway links Grain County to two major metropolitan areas that are within seventy-five miles, but there is little commuting for employment.

Model County is highly dependent on agriculture and related services, since nonfarm industry is underdeveloped in the county. Like Grain County, Model County is linked by an interstate highway with two metropolitan areas and a large city, but here there is considerable commuting for employment. However, unlike Grain County farms, which tend to be large and highly specialized in cash crops, Model County farms are smaller than the state average and very diversified. Tobacco, some small crops of cash grains, hay, and all types of livestock farming, including small dairy operations, are found in the county. In fact, livestock generates more market value of products sold than do crops. Farmers in Model County have also diversified into vegetables and other, more exotic commodities, such as trout and rabbits. In many ways Model County farms represent the idealized image of the small family farm.

Dairy County's agricultural base is indicated by its name. It is one of the top ten dairy counties in the state. Most dairy farmers also raise tobacco, which is a highly profitable crop that can be raised around the daily milking schedule and in fact is the leading cash crop in the county. Hay for use by local dairy and beef farms and for commercial sales has the greatest acreage base. These are the only crops of significance. Although most people think of dairying as the primary agricultural pursuit in the community, there are actually more beef cattle on hand, and sheep production is making a major comeback after three decades of decline. Farms in Dairy County, therefore, are neither as specialized as those in Grain County nor as diversified as those in Model County. Agriculture is the primary industry in Dairy County, with nearly one in four workers directly involved in agriculture or related firms. Other local private firms provide employment for nearly one in three of the county's workers, and the others find employment elsewhere, although commuting for employment is more difficult, since there is no four-lane highway crossing the county.

Grain, Model, and Dairy Counties reflect the diversity of agriculturally dependent communities in America. Farming contributes a significant proportion of the total personal income in each county, whereas opportunities for off-farm employment vary depending on historical patterns of development and the happenstance of geography. Local traditions and family histories are tied to agriculture, and most residents are involved in farming either directly or through supporting services or are only one generation away from the farm.

"Basically, Small Town is tobacco barns. Now in Model Town they have one factory, and there are some new subdivisions. But a lot of the shops have closed, I'm sure because of the problems that farmers have had in the last couple of years. This is a farming community; if the farmers don't have money to spend, then the shops are not going to stay open." Rural farm communities have always been affected by economic crises in farming. During the Dust Bowl years, rural communities throughout the Midwest were devastated by the collapse of farming. As farming mechanized in the South, the demand for labor declined, and a major migration of African Americans began. Each technological innovation that has changed the nature and scale of farming has had ripple effects on rural communities, as has each farm financial crisis. In many cases these effects have been compounded by larger social and economic trends in society. For example, the out-migration of African Americans from the shrinking job market of the agricultural areas of the South was encouraged by the rapid economic growth of the industrial cities of the North, which offered new job opportunities to potential migrants.

When farmers are worried about their finances, business owners in rural communities worry too. The business of farmers is production, but the farm enterprise and the farm family are also consumers. In order to produce agricultural products, farmers must purchase inputs, and in order to live, farm families must buy consumer goods and services. The economic ties are close: the businesses of rural farm communities thrive or suffer along with the farmers.

In time of economic recession, especially hard hit are the businesses that supply inputs for the farm operations. We studied agricultural cooperatives in the early '80s, and the managers of these supply stores explained that they no longer could afford to allow farmers to purchase seeds and chemicals in the spring for payment in the fall when the crop is sold. Too many large, successful farmers suddenly found themselves deeply in debt or bankrupt, and the farm supply cooperatives faced major losses from bad debts. Although they regretted having to change their policies, they also had to make a business decision that would protect their consumer investors. The days of buying on credit have pretty well ended in farming America.

Although farmers cannot put off buying the seeds and chemicals they need to put out a crop or the feed they need for their livestock, they can delay purchasing big-ticket items like tractors, combines, and harvesters. One can always keep repairing what one has or continue working with equipment that is not big enough or right for the job, instead of buying something new. Therefore, the implement dealers are often the first to experience sympathetic economic pains with farmers. "Grain Town has got two dealers, and neither one of them is very big, and neither is able to keep the equipment people need. So if I need a part, I've got to go to one of the out-of-town dealers, or go through these people up here and let them order my parts and wait two or three days for the parts to come in. Of course, those people are just like farmers, they thought big and bought big and invested big and went broke. Big equipment is hard on dealers too. They've got to invest so much and then they have to sell it so close that they can't afford to service it. It cost so much to service it. That's what the dealer that went broke told me."

The studies of hard-hit midwestern farm communities note other kinds of effects from a crisis in farming. Farm families often delay making time payments on consumer or farm goods, thereby affecting the ability of businesses to maintain their inventories. Mortgage payments are delayed, affecting the solvency of local lenders, and many also delay property tax payments, shrinking the income of local governments. The effect on local governments is intensified by a decline in land values, which reduces the local tax base. Finally, as more bankruptcies occur among farm families, many chose to leave the community. Others, realizing that their debt load is so large that they may never climb out, decide to leave farming with as much of their invested capital as possible. And so farms come up for sale. Sometimes they become new subdivisions, sometimes they are bought or rented by other farmers and stay in agricultural production, and sometimes the land just lies idle. Waiting.

Another ripple effect of the farm crisis for rural communities can be seen in the churches. It is not a crisis of faith, but simply a lack of population for local congregations. The decision of many young persons not to enter farming and the loss of young farm families who have not been able to financially survive the cycles of boom and bust have significantly reduced the number of young persons in farm-dependent communities. The churches, which have been the heart of community life, have suffered.

One young farm couple in their late twenties had started a dairy farm in Model County. They noted that the next youngest farm family in the area were in their late thirties. This couple described what it was like the first time they attended their local church. They were running late, and so the service had already started when they arrived. When they opened the doors to enter, "It was a sea of white-haired heads, and I said to my wife, 'I guess we can forget about child care during the service.'" Many other farm families commented on their declining church memberships and their concerns for the future of their churches. As another farmer in his early sixties explained, "We don't have many young people in our church. We have only one boy and his wife. We're all getting older and the young people are not there any more."

The losses in congregations have been large enough that many churches can no longer support a minister. Sharing ministers by two and sometimes three churches is becoming more common, and as a result the traditional roles of the minister—as spiritual shepherd and social leader—in the community are diminished by the need to travel from one church to another. And the traditional roles of the church—as the center of social life and the heart of community support—are lost with the declining membership. In some rural communities, like Hardin in Breckinridge County, new people are moving in from more distant urban centers, but they often bring a different understanding of the role of the church in their lives and in the life of their community, and sometimes they do not join existing churches. In many small rural communities, then, the ripple effects of the farm crises on the local churches represent the most significant consequence, because the church is often the cornerstone of community life.

What is happening to the church membership in many farm communities is just another symptom of the interdependence of the nature of rural life and the vitality of the farm sector. And it is a model of how rural population loss has made it more and more difficult for rural communities to sustain other professionals, such as doctors, dentists, and accountants. Like the ripples that spread from a rock thrown into a pond, all aspects of life in a farming-dependent community are touched when farm families experience a financial crisis. The longer and more serious the economic crisis, the more enduring the consequences for the related communities.

But there are other forces working on rural communities besides the changes and episodic crises in agriculture. The urbanization of society is continuing. Urban-owned businesses are reaching out into the rural hinterland with regional shopping malls and discount retail stores. Better roads have reduced the travel time to urban retail centers and made it easier for an urban worker to find a home in the country and commute to work. The Sears mail-order catalog once was a staple of rural households, but even this has become a part of history, for the nature of the mail-order retail market is changing. There is more diversity in the products offered, and even services are available by mail order, extending the competition for the consumer dollar beyond state lines. Changes in state and federal laws have encouraged the acquisition of rural financial institutions by larger, urban-based ones, contributing in turn to other changes that are not rooted in the economic troubles of the farm sector.

The changes in rural communities can be understood with reference to two major factors, the distance of the community from a large urban center and the size of the community population. Those rural communities, regardless of size, that are closest to the urban centers have been most absorbed into the growing sphere of urban influence. Among the more distant rural communities, the commercial and business lives of the smallest ones have been absorbed by the larger ones that have succeeded in diversifying their economies. Although this is a simplified explanation of the very complex workings of urbanization and economic expansion, it does capture the larger processes of change that are reshaping rural America.

The most visible change has been on rural main streets. In most rural communities Main Street has been the hub of commercial and social life. Main Street was bustling with many small, locally owned and operated businesses and services. Nearly all of a family's consumer needs could be satisfied, and most of the stores provided a wide variety of consumer goods. In difficult times store owners would be willing to let customers purchase on time, because they were neighbors as well as customers. If a customer wanted something the Main Street store did not have, it would be ordered and waiting for the next trip to town. "Look at Agtown. Five or seven years ago it had a whole lot of businesses, more so than it's got now. It was a little small town, only about two hundred people living there, but it was just like a big town. It had everything. It had a bank, city hall, police department, a shoe store, a car dealer, a hardware store, and a paint store. They had dry cleaning places, they had a laundromat and a whole bunch of different little businesses. They're all gone. This has been in the last five years. The only thing that is left is an auto part store. The bank is still in business, but I would think that in the next three or four years it will be bought by somebody. I don't see how it can survive the way it is going. The only thing doing good business is the new convenient store."

Main Street stores of rural communities near urban centers cannot compete with regional malls nor the easier access to the retail outlets of cities that now are closer because of intrastate highway systems. Moreover, national discounters such as Wal-Mart have targeted the larger rural communities as their primary marketing strategy. These competitors have changed the role of Main Street businesses from providers of all retail needs to providers of specialized low cost, high volume consumer goods such as groceries, pharmaceutical products, hardware supplies, and more recently, videos. Many of the other traditional businesses (for example, clothing and furniture) have closed forever or left the smaller rural towns for larger ones. "It used to be on Saturday afternoon everybody went to town unless there was just something on the farm they couldn't leave. It used to be that if you went to Model Town and wanted to park anywhere close to the business places you had to go early. Otherwise there would be no parking places. When you got there the sidewalks were so full you were rubbing elbows. But now in our little country town, you can go mid–Saturday afternoon and there is less cars than anytime in the week because the courthouse is closed. There is not much that goes on any more."

A popular myth about farming, based in fact in the past, is that farm families are self-sufficient: they raise their own meat and vegetables and so do not need as much to live on as do urban families. However, farmers are no longer their own food store. Farms with some chickens, a dairy cow, a few hogs, and a large garden are no longer common. Farm families buy their groceries and other consumer goods just as any other family does. But the smaller, more distant rural villages have seen a continued decline in the number and diversity of locally owned stores, and those that remain no longer offer the variety of products and services that the farmers' parents once found. "There used to be six or seven little country stores around here. There was competition. You learned to make the circuit to see if you could pay a few cents less for a dozen eggs than what they were asking over here. If a box of matches went for five cents a box or a penny a box on up the line, you'd find out about those things. Now there's only one or two little stores left, and maybe we buy something there if we run out during the week. But we go to the big grocery store in Metro City once a week for most of our groceries."

The locally owned stores have been replaced by franchises and chain stores. The Mom and Pop store has been replaced by the SuperAmerica and the Minit Mart. The home cooking restaurant has been replaced by the Dairy Queen or the Burger King. The new stores are all locally managed, but under business guidelines set up by the home office and with a percentage of profits leaving the local community. The consumer dollars that do circulate in rural communities no longer have the same impact as they once did.

The small family-owned businesses are like farm families, caught in a cost-price squeeze. The surviving small-town retail stores must pay higher costs to provide their customers goods. The long distances from urban wholesale centers and the smaller customer demand in rural communities means these retailers cannot take advantage of the cost savings that come from volume buying. And, like farm families, finding labor at a cost they can afford is difficult, because they are competing with larger stores for good labor. As a result even when these businesses offer the same products as the chain stores, their prices are higher. And like smart shoppers everywhere, farm families look for a bargain. "It is kind of sad, but I guess in a way we are to blame. You can get things so much cheaper at Small Town. You can get more stuff there than you can at these little country stores, and it's not as high. The little stores around here are high.

They've got to set the prices up to try and stay in business, I reckon. There are about seven or eight big huge store buildings sitting there empty that used to be thriving."

The changes in the retail bases of nearby towns means that often it is no longer possible to simply "run into town" for a few things. Many consumer goods can be found only by traveling to larger rural communities or regional retail centers. As the costs of items in local stores have risen, all rural families must make the choice between paying higher prices locally or traveling elsewhere. The improved road system makes it nearly as easy to travel twenty miles as it once was to travel two or three.

Along with the loss of retail stores and other business services have come other changes. Some rural communities, especially the smaller ones most distant from cities, have faced the gradual loss of population, because they are less likely to have competed successfully for new businesses. Younger people have moved elsewhere looking for employment opportunities, and the population has aged. In other rural communities, new industries and new roads have brought new people who have no ties to farming. The cumulative effects of these economic and population changes have altered social life.

In rural communities closest to large cities, residential growth is occurring, putting pressure on land values and making it difficult for farmers to compete for the best land. Newcomers also demand more and better public services, which often lead to higher property taxes, increasing the financial pressures on farm families. Other rural communities have succeeded in diversifying their local economies, attracting manufacturing plants, government services, or other types of industry. The development of new industries in rural communities is a blessing for long-time residents, including farm family members, who have been searching for new or better job opportunities. New industries serve to anchor young adults to the community. Instead of watching their children graduate from high school and go off to college, or in search of job opportunities, perhaps never to return, the community now offers real jobs with wages often competitive to those in cities, and more of the young people remain in the area. But new industries also mean new residents, many with no ties to farming.

In these growing rural communities, social life has changed. Sometimes the changes are so small that most nonfarmers would not notice. Community organizations hold meetings at times when most farmers are in the fields, or events that once celebrated the importance of agriculture are forgotten or simply not well attended. Sometimes the changes in the community are marked more by what is no longer there than by something new, as seen in how this farmer describes his trip to town today. "How's the community changing? Well, of course, the big thing is like I said, we used to all be farmers and everybody depended on their living for farming. Now, you go to town and you're not going to see many fences."

In declining communities, on the other hand, social activities become more difficult to sustain, because the people simply are no longer there. The church-sponsored softball team no longer plays in tournaments because it has difficulty finding enough people to play. The local elementary school is consolidated with the one in a neighboring town because there just are not enough children to justify the costs of keeping the local school open. The annual Christmas festival is canceled because there are not enough volunteers to organize and operate the booths.

The loss of locally owned businesses, the arrival of nonfarm residents, the diversification of the local economy, and the loss of operating farms all contribute to a changing sense of the meaning of the community. For farmers and many other rural residents, the changes that have been occurring in their communities signal a change of identity, a change in their sense of place in the larger social world. The physical and population changes are visible reminders that not only has farming as a business changed, but the world in which they and their families live out their daily lives is also changing.

In most cases a farm family is tied to a particular place because of the nature of their work, and their local community becomes the key focus of their daily lives. Of course, most families' lives are bound to a locality, regardless of where they live or the work they do. Our communities are where we work, shop, worship, and play. Our communities have special meanings for us; urban residents are quick to correct others who say they live in a city when in fact they live in one of the suburbs. In communities across the nation, citizen groups organize to fight development proposals that they believe will change the character of where they live. How we define our community is important, for it is part and parcel of who we are, the kinds of neighbors we have, and the nature of our immediate social world. The economic, population, and social changes occurring in rural communities has been forcing farm families to rethink the meaning of their communities. "How has the town changed? It is not as farm-oriented as it used to be, because we've got a lot of factories that have come in. It used to be that even the people that lived in town knew what farming was all about because they had moved there from the farm. But it's not like that any more."

Sometimes the little things in life are what make us realize that our communities are changing. It is true, as the farmer said earlier, that you do not see as many fences when you drive to town and that there are "for sale" signs on your neighbors' farms. It is true that the pace and nature of life is changing. And it is true that there are fewer or no farm stores when you drive into town, and you have to travel somewhere else to find your farm supplies. No matter what causes you to stop short and realize that things are different now, the surprise and sense of loss are real. "Even the small country stores, they used to stay open until ten or eleven o'clock. But any more, all the country stores will close before we even get through working. They'll close at five when we're usually not done milking until eight or nine o'clock. And if you break anything after five o'clock, you have to wait until the next day because nothing in town is open. They've got their own schedule, and it isn't a schedule the farmer keeps."

These changes frequently signal a new position for farmers in their communities. In years past, farm families were truly the driving force. The Farm Bureau and other farm organizations sponsored community activities and youth groups. Farmers served on the boards of voluntary associations and other community organizations. One farmer recalled that the people who lived outside Grain Town had public water systems because the farmers had organized to pay for laying the pipes and set up the local water districts. But as their communities change and the pressures to stay afloat financially increase, demanding greater time commitments to the farm enterprise and to off-farm jobs, farmers find they do not have the time to participate as much as they once did. There is a sense that the respect that was once accorded to farmers has declined with the declining importance of their businesses to the economic life of the community. And farmers see their position in the formal and informal political system of the community declining as they are replaced by new leaders drawn from new businesses and the new nonfarm population. Hence, the shift in community identity from a farming community to something else also signals a change in the social standing of farmers.

This shift in community identity is hastened by in-migration. The new residents own businesses or have jobs in nearby cities. They have the interest and, unlike farm families, often the time, to participate in community affairs. Studies of rural communities experiencing population growth from in-migration suggest that conflict often arises when newcomers begin to exert influence over local affairs and their expectations for services and activities collide with customary ways of doing things. In a sense there are mixed blessings associated with newcomers. They bring new money, new businesses, and new families into rural communities, but they also bring new ideas, new expectations, and new ways of life.

The community no longer revolves around and sometimes no longer has regard for farming, and farm families become just one other interest group in the community. "That's one thing I don't think the city of Dairy Town realizes either, that farming is still important. All the City Council does is run a two-page ad in the paper every year during Dairy Month that says, 'We salute you.' That's about it. The majority of the rest of the businesses ignore us."

Some of the most unneighborly conflicts erupt over the sale of farm lands for residential development. When developers begin to eye a parcel of land, the price goes up. On one hand, this is an opportunity for some, especially older, farmers to sell land at an urban-use price as a way to prepare for retirement, whereas other farmers see a chance to hedge against bad debts or unexpected health costs. On the other hand, there is a recognition of the problems that can develop when nonfarm neighbors complain about farm production practices or when their children are injured playing around equipment or other parts of the neighboring farm, such as a stock pond, because they do not understand the dangers. As Barb said, residential growth can make it more difficult for farmers to do business.

One aspect of farm and rural life that is both different and yet the same is the relationships among neighbors. Our collective images of rural community life have been shaped by family memories and the popular media. In Walnut Grove, the Ingalls family and their neighbors worked together to help each other succeed, and in the movie *Witness* the most vivid visual image was the communal barn-raising. Old pictures from yesterday show us families working with their neighbors to plant or bring in a crop, to put up a barn, or to brand the calves. Relatively isolated from the towns and cities, the need for neighbors to be able to rely upon each other, to know that when you call, they will be there, has seemed to be the essence of rural farm life. One farmer recalled the most essential assistance once offered by neighbors and how it has changed. "The way it used to be was if somebody died the neighbor went and dug the grave. Now the funeral home hires a backhoe to come out here and do it."

The neighboring of farm life is first built on providing assistance, and that assistance is based on the fact that in the physical isolation of rural areas, you develop an awareness that if you can call on your neighbors, they can call on you. The nature of farming in earlier decades added to this dependence. Prior to the extensive mechanization of farming, hard, back-breaking labor requiring many hands was what farming was about. "We used to chop silage when I was a kid. We'd go and fill all the silos around in the county. People just don't do that any more, they are more independent." Hand labor meant that if you were to get your crop out or harvest it, you had to have access to as many workers as possible. In most cases your own family could not provide all the labor you needed, so "swapping" work became an integral part of farming.

But the new machines changed this. At first only a few farmers could afford the new machines, and so farmers would share the machines, moving them from one farm to another in a rural neighborhood. "It is not like in the old days when they had the threshing machines. The thresher would go around from farm to farm, and everybody, all the neighbors, would go in and help each other. They don't do that any more. Of course, they don't thresh any more; they have combines now. Machinery has done a lot to change things." But as more and larger machines came onto the market their cost began to decline, so more and more farmers could afford their own machines. You now could work all your land on your own, without the help of your neighbors. And as farms got bigger, it actually became a liability to take time from working your land to help a neighbor.

"Ten or twenty years ago, everybody had a small farm and everybody worked together. As time went along we saw the operations become bigger, and we saw the closeness between the neighbors drift apart. They became a lot more competitive. We saw not only the farming operations themselves change, but we saw the farmers, too, change from being a close-knit group in the community to a lot more independent, 'me first' type. They're not helping each other as much. When you are a bigger business you get more independent, and you don't have the time to be neighborly when you get so big."

It is not that swapping work no longer occurs; it just occurs less often, and it is more difficult to arrange. The tight scheduling of production activities, the narrow profit margins, and the need for skilled labor to work the machinery all mean that swapping can be risky. Moreover, although you might be able to help a neighbor when you are needed, the neighbor might not be there for you—not because the neighbors do not want to help, but perhaps when you need them, they will have to be doing their own work on their own farms. "We don't swap work much any more. Every time I swap work I feel like I get burnt. Everybody is so busy with their own enterprise that they don't have time. There are times when you swap a little, but not like they used to."

The bonds of friendship were different when farmers realized that their success depended upon their neighbors' willingness to swap work or share equipment. In the relative isolation of farming communities, your neighbor's children used to be like your own, for the children spent their free time at each other's homes, and all the children helped in the farm work. As the pace and nature of farming changed, and as families' and children's expectations about their opportunities to participate in after-school and summer activities changed, children began to spend less time at home on the farm or with their neighbors. And another bond of friend-ship loosened. "There has been something lost in the way of life. There was this sense of family; you felt you were more close knit. What con-cerned your neighbor did concern you, and this was returned in full mea-sure. Now it is just not that way any more."

An essential part of neighboring yesterday and today is the casual conversations among friends. Although farming is a labor intensive occu-pation, and the pace of farm life is hectic, there have always been oppor-tunities for spontaneous and informal neighboring. "In the '50s and '60s I could be working out here in the field and neighbors would come by. We might have five of us jaw-jawing. You don't have that any more. I hate to lose that; it was a great joy in my life. Now if one comes by and throws his hand up, you're lucky. Nobody takes time to talk any more." Older farmers remember when the small country stores were the places where neighbors often gathered. The small country stores that remain still sell a limited amount of groceries, some hardware, and other small house-hold items. But their biggest sales are sandwiches, snacks, soft drinks, and cigarettes. Drive past any of these small rural stores early in the morn-ing or at noon, and the parking lots are filled with pickup trucks. Even in the winter, when farm work is slow, the habits of the summer continue, and farmers come to talk about the season past and the season to come. It is during these sessions that they learn about what is happening in the neighborhood, what new products there are, farming techniques, or cur-rent or proposed government programs. Such casual conversations are the oil of community life, especially in more isolated farm neighborhoods, where the opportunities for other kinds of entertainment are limited.

Some of the farm wives argue that despite stereotypes about women gossiping, their husbands are the ones who spend time talking with neigh-bors. Their husbands are more likely to be in the fields to see a neighbor passing by, to stop by the feed store, or to take a crop or livestock to mar-ket, and so have the opportunities to meet with neighbors. Or maybe a neighbor comes over to see whether you have a part or can help repair some machinery that has broken down, and he "sets a spell" to talk. The little country back roads serve as the physical telegraph linking neighbors who stop by to talk on the way to somewhere else. One farm wife de-scribed the informal communication network among the menfolk this way: "The men do the gossiping. Us women don't have time to go visit. If they are out working and see each other they will stop and they'll talk and talk and talk. It is the men that do the talking instead of the women. Myself, I don't have the time to go out to a neighbor's to visit. Sometimes they will stay out there a long time talking, in the garage, under the shade trees."

But even this, the heart of neighboring, is changing. It is not that people do not talk any more, just that things have changed. There is less time to visit, and it is not as easy as it used to be. As farms have become bigger, people's neighbors are farther away. A road that used to have five or six farmsteads on it now has only one. And sometimes the farm neighbors are gone, replaced by residential subdivisions filled with families whose interest in farming is only secondhand.

Furthermore, the highway has had an influence on neighboring. As we speed down our interstate or intrastate highways, we enjoy looking at the passing landscape and appreciate the time we save traveling these roads. But highways, like Robert Frost's fences, separate people who once were neighbors. When major highways or roads are built, there is no accommodation made for maintaining the ease of movement from one side to another. If a farmer's property is divided, the state or federal government pays for the land that is taken in construction. But how the farmer continues to work a farm operation that is split in half, or how neighbors manage to maintain their customary visiting is not a concern of government. These roads change the patterns of life in ways that the designers, the builders, and those that travel them, do not realize. "When the highway came through it changed this whole neighborhood. It was in '65 when they built the thing, and it has never been the same since. It cut up farms and it cut up neighbors. It just cut the heart out of us when that road came through here."

New roads through the community are appreciated by farm families for the economic growth they bring and the ease of travel. But there is also a recognition that these benefits are not without costs. Public roads are a public good, and the few have to sacrifice for the good of the many: that is the nature of our system of government. No change is without costs, and farm families sometimes feel that the costs are often ones that only they experience, and only they care about. "We're so mobile now. It used to be that if you broke a bolt or something like that you went over to the neighbors to see if he had one, and you'd borrow it. Or you went up here to the little country store to get it, and you'd sit a spell to talk. But now, we just hop in the car and go to town."

Despite the changes in their communities and in the fundamental way of doing their work, farm families still know, "If you ever need help in anything they come and help. Yeah. If I need help, then I call on my neighbors and they will come, and I would do the same thing for them." The sense of a community of supportive neighbors goes beyond the physical boundaries of where they live. Farmers, regardless of their location, the commodities they raise, or their financial condition, have a common sense of struggle for survival. It is because of that universal bond that farm families understand the sense of helplessness and despair felt by those experiencing a drought, for example, and offer to help, although they may have never met each other. "Last year as a result of the drought in North Carolina, a lot of people were saved because of the haylift that came in from the Midwest. Now it is right the opposite. North Carolina is sending truckloads of hay up here. I was surprised at it, in a way, because I didn't think that farmers could be that generous now. This is what we heard about thirty years ago, farmers helping each other. But I guess I was a little bit surprised last year that so many people would sacrifice that much to do something for somebody three or four states away."

The "hay lifts" during the droughts of the late '80s had some government involvement, but they relied essentially on the generosity of farmers who donated hay, truckers who volunteered their rigs and their time, and railroad companies who donated flatbeds and rail space. Local Cooperative Extension Service agents helped organize the effort and the ASCS (Agricultural Stabilization and Conservation Service) coordinated the distribution of donated hay. It was a grand example of mutual assistance within a community of farm families, and the crisis reminded farmers that in hard times they must rely upon each other.

In some ways farmers feel that their familiar world is slipping away from them. Larger forces in the society are changing their communities, altering their relationships with their neighbors, and marginalizing their economic and social roles in their communities' lives, while also providing new opportunities to increase their family incomes. The neighbors who once shared a livelihood and a lifestyle with them have either left for the cities or have been replaced by families interested only in a rural lifestyle and not necessarily in farming.

Yet, in the face of all these changes, certain elements of life in farm communities remain the same. You can still count on your farm neighbor to help in times of need, and there is still a deep-rooted belief that life in a smaller community offers a better place for raising a family than anywhere else. And, although farmers sometimes feel bewildered and resentful about the changes going on around them, there is also a recognition that these changes have also made certain aspects of life easier.

As in all the other facets of their lives, farm families recognize that changes have occurred and will continue to occur in their communities and in their relationships with their neighbors. They also know that these changes have altered the terms of their lives. And they know that their challenge is to adjust without losing what makes farming worthwhile as a business and a way of life.

9 An Unbroken Circle of Hope

BOUQUET

At 1:15 I awaken,
her son breathing softly beside me.
He handles it well by day
but at night his body constricts
ever so often as if he is battling
    the cancer that overtakes her
    daily.

and I am awake, wondering
what kind of night is she having
and did her headache subside and
is she awake wondering
    where is the headache coming from
    and will her daughter arrive
in time? *Does she know*

I have always admired
    the way she sets a table
    the plants that flourish
        inside and outside her home
    her sheer delight in little ones
    the utter sense of family
        that is her mantle

that I would bring my babies here again
    to grow up on this farm? Tomorrow
I will look for something at the florist's
a rose for her room
the dark eyes, the wide smile
the long tapering hands that have tended
hens and herbs and thirteen grandchildren

a rose for Mary's days.

BASIN SPRING FARM

Jim's mother died on March 13, 1991. Mother Mary.

I cannot believe that she is gone. I cannot imagine that we will never again fix a Sunday meal together and put all the leftovers up for supper, or cut corn off the cob for the freezer, or ride in the truck out to the field to see how the men are doing, or leaf through old photographs while she tells me of this relative or that one.

I will no longer enjoy her banana cake with brown sugar icing, or the best fried chicken this side of the Mason-Dixon line with her lighter-than-air baking soda biscuits, or delight in a warm slice of her chocolate skillet pie on a gray January afternoon. And I will never again hear her call me "Bar-bara-Allen," the nickname she gave me years ago from the folk ballad.

It was lung cancer. She was diagnosed in early September of 1990 and had surgery shortly thereafter to remove a tumor from her left lung. A tumor in the other lung was left alone, because the two operations would have been too hard on her. The doctors decided to treat both lungs with radia-tion after she recovered from the surgery, with the hope of arresting the disease in both sites.

She recovered beautifully from surgery and received thirty-three radia-tion treatments that fall. Several family members helped with driving to the radiation center in Louisville, but most often it was my father-in-law who drove her—five days a week for two weeks, with a week off between cycles. She finished the treatments just before Christmas. We decided to forgo the fam-ily open house between Christmas and New Year's that year, which usually included her country ham, pecan pies, jam cake, or rum cake. She snuggled into a Christmas sweatshirt I gave her that read "I Love You Deerly" and spent a quiet holiday reading and resting. She was tired. We were all tired, but we were glad to be together that season, not knowing that it would be our last together.

Just after the first of the year she was back in the hospital again. The news was grave. The cancer had spread throughout her skeletal system, and the internist gave her two weeks to two months to live.

The trips to Louisville for radiation treatments resumed to keep her mo-bile and to keep her body functioning for as along as possible while the disease ran its course. On the eighth of March, however, the cancer metas-tasized to her brain, the same afternoon that Hospice arrived. For five days and five nights her daughters, Sue and Laura, her sister, Jessie, her grandson's wife Marsha—a true "angel of mercy" who showed us the fine art of nursing—and I ministered to her needs under the guidance of Hos-pice, with various members of the family.

Nearly two months to the day from her final diagnosis, Mary Fontaine Scott Foote died at home in her bedroom overlooking Sinking Creek, family at hand, with loving handwork of Aunt Jessie' Tree of Life quilt spread across the bed. She was buried on Saint Patrick's day, a very March day of wind and rain, in her native Meade County near the graves of her parents, Walter and Ruth Scott, and her sister Rena Lou.

Her death touched me as death never had. It marked the end of an era, an era that began nearly twenty-four Easter Sundays ago, when we met for the first time at Basin Spring. There was a large gathering of sons and daughters and grandchildren for her funeral, and scores of neighbors and friends who joined us for the service and the long procession from Irvington to the Cap Anderson Cemetery in Meade county.

We had learned a lot about Mary Fontaine Scott Foote during her last days of grace and resolve, and we learned a lot about ourselves and one other. And I felt for the first time a shift in the fault line between generations. It occurred on the morning of the funeral. Marianna, our niece, was staying here with a guest, and we were all getting ready for the service when she remarked, "If you ever want to get rid of this house, I want you to let me know."

Surely I had not heard her correctly. This was the house that had plagued Mary Foote for more than 30 years, and me for nearly 14 with its 120-year-old idiosyncrasies: four gargantuan chimneys set inside the roofline, which guaranteed flashing disasters sooner if not later without frequent maintenance; wood heat (with a gas backup) downstairs; no heat upstairs; air conditioning au naturel in the summer via ceiling fans and open windows; and assorted critters of the four-legged and eight-legged varieties common to old houses in the country. Some of its quirks of character I had come to treasure; others were merely endured in the cranky old dear of a house that everyone loved but nobody wanted. No one but us, because we wanted Basin Spring. These thoughts ran through my mind as I regarded my niece.

"You have got to be kidding," I told her.

"No," she said, "I mean it."

The rest of the conversation was lost to me, something about "so many memories," as I realized what her words conveyed.

Within hours we had entered a new dimension, our perspectives changed forever. Though Basin Spring remained very much the same, her generations were being pulled hand in hand through time.

I thought of the ranch wife I had met in 1978 at the First National Conference of Rural American Women in Washington, D.C., and later at the First White House Conference of Farm Women in 1980. I had seen her last at the Second National Conference of Farm Women in Historical Perspective at the University of Wisconsin in 1986. She and her husband had run a ranch in Colorado for twenty-six years and raised a family there. Spouse, mother, and rancher, she had also served as the Colorado president of Women Involved in Farm Economics (WIFE) and was at the time serving a two-year term as national president. Her words at the conference were seared into my memory:

"We don't want our sons to take over the ranch," she had said firmly. "We are discouraging them. What we have gone through is too hard." Her face was set. I understood now the weariness behind her words, the resolve of more than twenty years on the land. Yet her words were a sad indictment on the future of agriculture in general. I understood her perspective. But here was my young niece saying that this life that we had tried to protect and extend when we moved to Basin Spring in the fall of 1974 held meaning for her, too. I heard her words and I was stunned.

Jim and I have never assumed that our children—or any family member—would pick up where we leave off at Basin Spring. Such an undertaking cannot be asked, let alone expected, because farming has grown into more than the labor of love that once sustained farming families. Technology, chemistry, global politics, and world markets have stretched the traditionally complex skills of land stewardship into a discipline that sometimes bears little resemblance to traditional farming. It takes more than a strong back to make it. You need all the wit, wisdom, perfect health, and heart that you can muster. And given these things in abundance, you may still need one or two off-farm incomes.

And yet here was my niece telling me on the morning of her grandmother's funeral that what we had worked for—and Mary and Gerard Foote before us, and Fan and Edwin Foote before them—was of value beyond our wanting it. It mattered to her.

I felt something shift inside me as I dressed to join the funeral gathering of generations for whom Mary Fontaine Scott Foote had been sister and aunt and mother and grandmother. The dream that Jim Foote had carried as a young graduate student had found its mark—to take care of "a corner of the earth"—when the rest of his generation (myself included) wanted nothing less than to stop a war and save the world. This farm would survive. The heritage and home and refuge for us all would last, somehow, because someone after Jim and me would love it. Marianna loved it. Our children loved it, too. And out of that generation, someone would extend a hand, and through sweat and pain and sweet ideal, Basin Spring would be pulled into a new era.

It must happen. It would happen. Because farmers are storytellers. Because . . .

IF YOU HAVE TO ASK WHY, YOU WOULDN'T UNDERSTAND THE ANSWER ANYWAY

During one of the interviews conducted for the study on which this book is based, we sat at the kitchen table sipping lemonade and talking with a farm couple in their early fifties one July morning. We had driven down a long, twisting gravel drive to a picture-postcard house set halfway down the hill. From any window you could look out and see the rolling hills with the cattle grazing and the thick forest that ringed the pastures. It had rained the night before, the first time in nearly fifty days. When we arrived the small talk focused on the drought, and we asked if it had rained there. "Yes." he said, with a sad little smile, and that was all. Not the response we expected to hear from a farmer after a rain finally broke another of the "worst droughts in history." But later we would come to understand his answer. As the conversation moved from how they had started in farming to what they saw as the advantages and disadvantages of raising children on a farm to their views on farming as a business, they began to talk about how it used to be. They had once felt they could not get ahead; now they knew they could not stay even.

This was their third bad year in a row. Last year they had had to borrow money to pay the interest on their farm loan. The drought had been so bad that the pastures had burned out completely. In the last two weeks they had stopped walking out in the fields because all they could hear was their cattle lowing from hunger. The "emergency hay lift" had begun two weeks earlier, and, desperate for feed for their cattle, they had borrowed against their tobacco crop to buy some of this hay. Then the rain had come the night before. At dusk the blessed clouds had rolled in from the south, coming over the hill, not to kiss but to pelt his farm with a couple of inches of rain and hail in less than half an hour. He figured he had lost almost half his tobacco. After they paid off the loan for the emergency hay they had borrowed against the tobacco, there was not going to be enough to pay the interest on the farm loan, and they would have to borrow the interest payment again. They both began talking about their dreams for the farm, their children, retiring someday, and then he ducked his head as a tear rolled down his cheek. We picked up our glasses of lemonade and his wife offered us some cookies. It was quiet for a while, and then he told us a story about what the farm had looked like when they bought it and the work they had done to bring back the land. By the time we left, we were laughing at another of their stories about life on the farm.

One early September evening Lori sat in the living room with a farm couple. She had called for an interview, and the wife had said to just come

by for dinner and there would be time for a comfortable talk. He had a
very good full-time job off the farm. They lived on and operated the farm
his father and grandfather had owned. She was a town girl and talked
about the shock that had come with their purchase of a combine and the
anger and frustration she had felt during her first drought. He talked much
about the challenges of his job in town and about the different challenges
with farming and how hard it is to make a profit or break even in farming
these days. When asked if he had ever used money from his town job to
offset losses from the farm, he allowed that he did not like doing it, but
that he had several times. Lori commented that people who did not farm
would probably find it hard to understand how someone could keep putting money into something that lost money as often as it made money.
"Why do you do it?" Lori asked. His wife leaned forward on the sofa and
gave a little laugh and said: "I can't wait to hear how he answers this!"
Her husband gave a rueful grin and said, "I fell into this one, didn't I?"
Clearly the question had struck a sensitive chord.

After leaving each interview, we would turn on the tape recorder and
just talk to each other about our impressions. What were the new ideas or
the threads of earlier interviews that had come up in the conversations,
and what would someone who did not farm find interesting about this
conversation? Did anything reflect or belie our common myths about farming and farm life? We began to realize that time after time we would leave
asking ourselves, Why do they go on? In spite of a common belief that
every year it is harder to make a living from farming, in spite of the difficulties in providing opportunities for children to participate in school
and organized recreational opportunities, and in spite of the uncertainty
that anyone would want the farm when the time comes, farm families could
rarely imagine anything besides retirement or death that would lead them
to quit farming.

We started to explore this in a roundabout fashion by asking what it
would take for the farm couples to quit farming. Sometimes surprised,
sometimes with a chuckle, the answers were always the same. They would
begin to talk about what farming or the farm means to them—the challenge, the sense of accomplishment and pride, the family history of the
place, the heritage they were building for their children, or the importance of being stewards of the land. They never really answered the question. Instead they spoke to what bound them to the farm and farming.

One of the debates about farming revolves around the question, Is
farming a business or is farming a lifestyle? One wonders if anyone thinks
that the answer to this simple question will somehow explain this unique
activity that is woven so tightly into our heritage. But what Barb, her niece,

and so many other farm families are saying is that this is the wrong question. It cannot be answered in words that would have meaning to those who have not shared the experience. This is not arrogance. It is an acknowledgment that the farming business and the farming lifestyle are intertwined and inseparable—in fact, and in farm families' hearts and minds. The myth that has surrounded farming is that somehow you can separate the business from the lifestyle, that you can dissect and analyze each and truly come to understand what each is about without considering the other.

So there is no answer to the question, Why go on in the face of all the risks and heartaches?—because as far as most farm families are concerned, that is not the question. The real question is, How can they quit a life that is the sum of their dreams and hopes, and a business that offers them challenges and the opportunity to be an independent entrepreneur? The following are stories of particular families we interviewed, but they represent the collective answer to this question. We refer to them by allegorical names.

NEWCOMERS TO FARMING—BUILDING NEW TRADITIONS

There are many ways to get into farming. There are some couples with no family background or roots in farming who choose farming as a career and a lifestyle. In recent years these couples have often been called back-to-landers, and to a great extent these newcomers to farming have been dismissed from any serious study, perhaps because they have been small-scale farmers who typically have chosen alternative commodities and practices. Yet these newcomers have something to tell us about why and how people choose farming as an economic activity and a lifestyle, and why they, like their multigeneration farm neighbors, persist. In Model County we interviewed three such couples, and two of their stories illustrate the power of the commitment to a way of life and a way to make a living.

The Newfarmers have lived in Model County for nearly eight years and have approximately sixty-five acres, thirty-eight on the original farm and the remaining acres on a recently purchased farm. They raise sheep and have a flock of about sixty head and a milk cow they use to raise dairy calves; they are planning to raise hay on their new acreage. She worked off the farm until the flock became large enough to warrant someone on the farm full-time. She weaves woolen mittens from the wool of their own sheep and plans to expand her home business and market her products outside the state. He has worked at several jobs and is currently employed full-time in a manufacturing plant. Neither of them has any background in farming, although the husband used to visit farm cousins. He attributes his early interest in farming to these visits, where "the kids seemed to have more things to do and seemed to be more content than I was in the suburbs."

The Newcomers have been in Model County for two and a half years and have recently moved to a forty-acre farm in a hilly portion of the county. They too raise sheep, but have a small flock of about twenty head. He has two small home-based businesses that provide the bulk of their income. She assists him in one, but spends most of the day working in the family garden and home-schooling their child. He was raised in a city and had no contact with farming while growing up. Her family owned and operated a very large dairy, but they did not live on the farm. Her only contact with the farm operation was, as a teen, milking the cows on Sundays so the hired hand could have the day off. She hated doing this job and resented the fact that this was how she got her allowance, "smelling like manure," whereas her friends were simply given their allowance. "I never envisioned that I would ever want to be involved with farming at all, yet here I am."

So what brought these two couples to a life on the land? Both couples attribute their decision to farm to an interest in ecology and alternative agricultural writers such as Wendell Berry amd Wes Jackson and magazines such as *Organic Farming*. The Newfarmers both subscribed to *Organic Gardening* and both read an article about a "learning" farm in the upper Midwest. Persons could spend six months to a year at the farm learning about agriculture—cropping, livestock, and machinery operation and maintenance. They met while attending this "school." He explains how they came to realize that they shared a common interest, one that could best be achieved together: "I really didn't have any full-time farming experience. When I read this article I felt like it was a good break for me. I had always wanted to get more into farming, but I didn't want to single-handedly go out and buy a farm and start farming. When I met my wife up there, it really was the center for both of us. Without each other, we wouldn't have had the gumption to go ahead and buy a place. It allowed us to really focus a little more."

The Newcomers met and married while in college, and their interest in farming arose from their classroom readings. Both recognized that, if they were to act on their dreams and their beliefs, they would have to gain some practical experience. So the Newcomers returned to her family's dairy, where for two years they did their "schooling" in farming. But their vision of farming was different from that of her parents, and so they moved to their own place. "We were reading people like Wendell Berry and Wes Jackson. People who farmed on a smaller scale than what I was accustomed to thinking from my parent's perspective, and were posing alternatives that appeal to my husband and I. Both of us want to live life off the earth. We don't want to strip away. We want to give back more than we are taking. Our view of farming is heavily philosophical. We realize that the earth is more than just farming. It is agriculture. There is a culture to the land. I think that is why we will stay with it. There is almost a religious quality to working with the land. I'm going to take care of my land just like I would a child. The land will take care of me as well. That would seem to be the thing that would scare me, if I had spent my whole life taking care of the land and in the end, the land did not take care of me. It would be just like being rejected by a mother or something. But we are practical about it, we have to make a living at it."

For both of these couples the decision to enter farming reflected a desire to make a living in a way that fit with their personal philosophies. Farming represented a way of life more environmentally harmonious than city life. Yet both approached this decision with careful planning. Both recognized the need to gain some experience in farming prior to beginning their own operations. And both carefully thought out the financial aspects of entering farming. Their selection of a place to settle, their choice of commodities, and their method of financing their farm all indicate a recognition that their lifestyle preference had to be balanced with realistic economic goals. Both developed financial plans that detailed how much they had to earn, both from farming and from off-farm jobs, over a five- or ten-year period, in order to pay for the land and still meet living expenses.

Although both have had income-generating employment off the farm, it was not enough to finance their dream. Like most farmers, they sought additional help through loans, but both couples discovered that no one was interested in making a loan to persons who had no background in farming. Although this is the Newfarmers explaining what happened to them, the Newcomers have a similar story: "At that time, because we had no previous farming experience that a bank would recognize, we weren't considered a good risk for a loan. They kind of laughed. 'You've got to be kidding,' one man said, 'I've been farming all my life and I don't make any money at it. How do you think you're going to make money at it?' The S&L where I had an account all my life said, 'If you want to buy a house we'll lend you the money.' They wouldn't lend me money for a farm. We told the man who owned the place how much we had to put down and how much we thought we could make a year. He agreed to that for five years, at which time he would need the balance. We felt that in five years we could easily refinance if we had a record of paying for five years and made improvements. But as it was, we paid it off in five years. But it wasn't the farm, it was both of us working off the farm."

Couples who have a background in farming have not encountered as many difficulties obtaining initial loans to begin their own operations. In a sense, the assumption of loan officers that long-term prior experience in farming is an essential eligibility requirement for obtaining a farm loan serves as a "gate" to the farming business that limits admission to a select few.

The acknowledgment that farming must be a business in order to sustain the lifestyle led these couples to make certain decisions about the type of farming they wanted to do. Their choices have been influenced by three factors: appropriate scale, diversity, and competitiveness. There are good reasons why, having come to farming from different points and experiences, both of these couples are raising sheep and have a mix of enterprises. First the Newfarmers and then the Newcomers explain why they chose to base their farm operations on sheep:

"It was a matter of reading about sheep and realizing there was something very appealing about working with them. First of all there is their size. You can deal with them more easily than cattle, and you don't need a vet for everything. They are cheap. You can actually get into it with very little money and make do with lots of things that cattle people would never be able to get by with. You can start small and build up a good healthy flock in just a few years. The gestation period for a lamb is five months, and it's nearly a year for cattle. Plus, it wouldn't put us in the same competitive market with other more experienced farmers."

"Talking with the county agent, he really thought that sheep were profitable on a small place; he encouraged us to think about sheep. We have always kind of had the feeling that we needed to be diversified and have a little of this and a little of that. I sell some eggs to people. With our cow's milk I make butter and sell a little butter. It doesn't pay a whole bunch, but just little things here and there. The chickens have paid for their feed through the sale of the eggs. If the chickens will pay for themselves, then their food is free. That is the way with our milk cow. We feel that, if the calf would pay for her feed and upkeep, then her milk is free to us."

Both couples believe that their scale of farming is the primary way for persons to enter this business, and both believe that their scale of farming is capable of providing them with an acceptable standard of living. Yet both have also recognized that in the early years the farm alone cannot provide for all their living expenses. The Newfarmers explain: "At this point, I don't know if we'll try and farm full-time. The farm has not generated real income. Every year we think we look good on paper, but we don't have any money in the bank. I feel like we can probably handle one hundred ewes on this farm. I think we could live on the money that they would generate."

Despite the reliance on off-farm income, both couples strongly assert their primary identity as farm families, because they are committed to making a living from the land. The Newcomers illustrate this perspective: "We are a farm. We consider it farming even though it doesn't count for a lot of our income, but it is some income. It is a great percentage for one or two months. It is very important to us. We are risking money and labor. I think that may be a redefinition of what farming actually is. Is there money at risk? How much? Is there substantial income for certain months? Tobacco people get a lot of money at certain times of the year. We get more money at certain times of the year from the sheep. The way you define a farm, in some definitions I've read, we're not even a blip on the scope."

Although these new entrants to farming are often dismissed as wide-eyed romantics who have no real understanding of farming, they share many things with more traditional farm families. Couples who entered farming in more customary ways—through growing up on a farm and with the assistance of relatives—would recognize as familiar these newcomers' descriptions of developing long-term financial plans with goals and objectives that would enable them to eventually use farming as their primary source of income. More traditional farm families would also recognize the frustration of drawing up detailed plans for the year and then ending up with no "money in the bank." Most farm families can recount similar difficulties in obtaining loans, many have sought to diversify their operations to make them more resistant to economic swings, and others would agree that small-scale farming is the best strategy for survival.

The nature of their farm operations and their lifestyles have also eased the newcomers' transition into the community. Model County once had large flocks of sheep, and both couples indicate that neighbors have been watching them and asking questions about raising sheep as an alternative source of farm income. Indeed, one couple hosted an open house so the neighbors who had been asking all the questions could come and see for themselves. Their work ethic, their love of the land, their willingness to build their farms slowly, all are ties between the new families and many of the older farm couples in their communities, as the Newfarmers explain: "We worked for a farmer in tobacco one year. When they found out how we had gotten our farm and what we were doing and building the house ourselves, they could empathize so much with what we were doing, because it is what they did when they first started out. They look at their children, who expect everything immediately or yesterday, if that were possible, who are heavily in debt, even though they are not on the farm, but just making a living. It is so different than what they had. They could really empathize and appreciate what we were doing."

These new farm families see themselves building a family tradition that they hope to pass on to their children. Despite the costs of entering into farming, the economic risks, and the low returns, they have committed themselves to a lifestyle and a business that has great value for them. Although these newcomers may express more clearly and openly their values and beliefs about the need to serve as stewards of the land and the importance of the culture of farming to society, many other farmers would echo their sentiments. Those outside farming might argue that these are sentimental views of farming that have no place in the modern business of farming and that these newcomers and the thousands of other small-scale farm families they are like are mere "blips on the scope" of national agricultural production. Yet these sentiments serve as a buffer to the harsh economic realities of farming and are the mortar that helps to build a lifelong commitment to farming.

STRIKING OUT ON THEIR OWN—CONTINUING THE TRADITION

The Dairykids had been operating a 279-acre dairy, corn, and hay farm for just about two years when we interviewed them. They were in their midtwenties and had two young children. He had spent many days on his grandfather's dairy farm while growing up; his father worked in a related agricultural field. The years he spent working with his grandfather and for other farmers convinced him that dairy farming was what he wanted to do. When he went to college he majored in animal sciences and worked on the school's dairy farm. After college he took a job working with a cattle operation and then moved to managing another university's dairy farm.

They met while in college, where she too majored in animal sciences. Growing up on a dairy farm had not "put her off farming" as it had many of her friends; rather, it had strengthened her desire to live the same life. After spending about a year managing the university dairy farm, they returned to her parents' farm to help with the family operation.

Her brother will take over the family's dairy. "My brother had been working there since he got out of school. He deserved it. He'd been the only one out of all of us that had put the time in it." While at her parents' farm the Dairykids worked with her father and brother, and their earnings were partly cash and partly in heifers. "The main reason we left the university farm was that all I got was a paycheck. Working for her dad, he set up a deal so that we could get part of our pay in heifers. That way we could start to gain equity, to get enough equity behind us to go out and be able to borrow money to buy a farm."

For a dairy farmer, the major cost of setting up the operation is the livestock; there is an overwhelming need to have a herd of young heifers able to produce sufficiently to generate the income necessary to pay for their feed and the rest of the operation. A good dairy cow is worth more than her weight in future profits. To receive your pay partly in heifers is one major step toward your own operation.

After working about two years for her father, they began searching for their own farm. It turned out to be a farm that had been idled through bankruptcy and then poorly managed when leased. Buildings were in disrepair, and every waterline in the house had broken. The land had suffered by not being cropped or properly managed—Johnson grass, thistles, chickweeds, "it was weeds competing with weeds." Although it was not ideal by any stretch of the imagination, it was possible that the farm could be had for a price that would fit into their financial plan. "We had worked it out on paper. Everything worked out, what we would need to borrow, how we would pay it back, figuring milk price and everything. We worked for weeks on those figures."

But when they took the financial plans to the local FmHA (Farmers Home Administration), they discovered that the tattered history of the farm they wanted to buy affected how the lenders looked at them. "He said, 'I don't want to talk to you about that place.'" They had lost a lot of money on the farm, and so had many local businesses, when the farmer who had been leasing the place left huge bills behind. For three weeks they talked to the FmHA, and one Friday the FmHA man said, "That's it, no more about the farm." But on Monday he called back and said he would sign the papers. The farm was theirs.

After the papers are signed, the real work begins. They are milking seventy-five cows and are preparing to begin milking the first crop of heifers they have raised. They are raising corn and alfalfa hay, because their financial edge depends on buying as little of their feed as possible. But the drought and the weeds are taking their toll, and production is not nearly what they needed to properly feed seventy-five cows. In the first year they did nearly all the work together, because they could not afford hired help. After their first child came, "we were just starting up, and she would take the baby down there in the carriage and she'd milk with me. We did that for a long time until we were just both wore out. Finally, we got so that we had a little bit of cash flow and was able to hire somebody to come help me milk." Today, besides caring for two children, she raises the calves, which requires hand feeding twice a day, and also serves as an extra hand at milking time.

Like the Newfarmers and the Newcomers, the Dairykids made a conscious choice to farm because, after weighing the advantages and disadvantages against the alternatives, they found the balance for farming favorable. After graduating from college, he was offered jobs in agricultural sales; he also had the opportunity to continue as a hired farm manager where his capital was not at risk. But what he wanted to do was to be his own boss in a situation in which he could build for their future.

Independent farming is an investment in the family's financial future, not in terms of great cash flows, but in terms of property and other capital. We asked if they would have made the same choice if they had not been so confident about being better off than those who chose another path. He said, "Well, I don't know." But his wife was much more certain. "Well, I think you would've. That's not one of the main things he looks at—Will I be better off than whoever? It's just what he's wanted to do. I think it's what he's always wanted to do." "Yeah, I never looked at it as what I was going to gain out of it but as what I wanted to do," he said. She added, "It's just like my dad always says, 'There's no money in farming. It's nothing but a way of life.' But I disagree with him a little bit on it that there's no money, because if you manage the things right and if you're lucky enough you'll live comfortably. You might not get rich. But he was right. I think, basically it's just a way of life. It's a way of living."

Although there is a focus on financial planning and investing for future security, the Dairykids also recognize that they are continuing a tradition. For her parents it is important that their oldest son has chosen to continue operating the family farm. "That place up there has been in the family for a hundred years. It's something you want to see continue." And her parents are proud that she has married someone who also wants to dairy. Although the Dairykids want their children to feel free to make their own life choices, they too hope that their new son will want to continue, that they can make their new farm into "the homeplace for ever and ever. He's going to hopefully be the legacy right there."

This young couple has made a commitment to a way of life that has special meaning and recognizes that they must overcome great economic challenges. But the drought of '88 has come at a difficult time. They have paid off what they borrowed for the farm, but the drought has forced them to borrow more operating capital than usual just to get through the summer. "The wells are getting dry. Things are getting tight. The worst is yet to come. The next year is going to be a tough one, with us just starting off." And if the rains do not come, they will have to continue borrowing to get through the winter. The financial plans worked out a few years ago are beginning to look like so much paper. But the optimism remains: "We are building something here."

Three years after this interview, we met someone from the area who knew the Dairykids and her family. He told us that the Dairykids had not been able to overcome the economic effects of two consecutive years of drought. They had been forced to sell their farm in order to keep some of what they had already invested. They had moved to another state, where he took a job as the manager of a large dairy. "Will they stay?" we asked. "For a while," he said, "but they are still hoping to have their own place someday." A dream deferred. A tradition postponed but not forgotten.

PASSING ON THE TRADITION

The Bridges operate a 202-acre farm and are in their early fifties. They raise corn, beans, wheat, and barley and feed most of it to their hogs, although with the purchase of some additional acreage they are selling more commercially. Both were raised on farms, their son is in partnership with them, and their siblings farm. They are the generation that bridges farming as it was yesterday and as it will be tomorrow. From the beginning he wanted to farm. "I definitely didn't want to do anything but farm. That was all I cared about. It was definitely farming that I wanted to do." For her, farming offered a way of life superior to its alternatives.

While they were growing up, both worked with their families on the family farms, he alongside his father and brothers in the fields, and she doing the small jobs that kept the household running and helping in the fields with the tobacco and the strawberries. They both remember a time when farming was labor intensive. "There was always a bunch of hands around. There would be ten or fifteen of us at times when we was in rushes. But most of the time there wasn't but four or five or six around." The work was hard, and when it was very hot people would sit under the shade trees during the afternoon and wait until it cooled down, and then go out and "do a whole day's work."

They started in farming the way many do: he and his father bought a farm together, and the young couple moved there after their marriage. She took a public job, part-time three days a week, shortly after they married, and still works part-time. At first she worked because they needed the money, but now she works because she enjoys being with other people and having interests outside the farm. While he continued to work the farm with his father, he did public work for ten years. It was not because he did not want to farm or because he particularly wanted a public job, but it was something new, an offer to take a struggling business and make it grow and share in the profits. "As it turned out it was one of the wisest moves I ever made. That's how we bought this place. I stayed in the public work until it got to the point and time that it was just more than I wanted to do. It put us in a financial position that we didn't have to worry about getting the farm fixed up where we wanted it."

Like most farm families, he and his brothers and his father continued working together, operating some farms in partnership and sharing equipment and labor with each other on their own farms. And now he and his son have bought the farm that he was born on, and they operate it together. Their son has a public job, helps with their farm, and also does custom work for other farmers in the community. Will the son stay in farming? The Bridges say that with the purchase of the original home place, even though it is only eighty-two acres, not big enough to support a family, he probably will. "There's a lot of history there. He's planning on living there. We're not sure but that's the way it looks now."

Although the Bridges were both raised on the farm, his son, like many young men his age, will marry a young woman from the country but not from farming. Many of the young women who have been raised on farms have no desire to marry back into farming. It is a hard life, there are more alternatives today than when their mothers were their age, and the risks are greater. The Bridges say that their son has tried to explain to his fiancée what farm life is like, "But I don't think she really realizes it."

Mr. Bridges shares a love for and commitment to farming with his brothers, but they differ in how they think about farming. It is a difference that has caused some tensions over the years, has led him to a different kind of farm operation, and has produced a different degree of economic security at this point in his life than for his brothers. "My family and I don't really agree. My wife and I, we don't want to be big. We just want to be comfortable and have a good living and not have to worry. When we bought something we looked at it from the point that we were to pay for it as quick as we could and get out of the debt and go on to another plateau. Now my dad, he liked big machinery and operating big. But he liked to pay for it first. My brothers don't look at it from that point of view, especially one of my brothers. He's in financial trouble because of it."

When they look around at many of the other farm families in their community, what they see affirms the value of their choice not to keep getting bigger and to remain diversified. Many of those in financial trouble are the ones "who wanted to be too big too fast, and they put all their eggs in one basket." But in addition to causing financial problems, being too big and too specialized can lead to poor farming practices, according to the Bridges, practices that result in soil erosion and exhausted land. The Bridges believe that if they care for the land, it will care for them. In the early '70s, following the rallying cry of government to plant fencerow to fencerow, the Bridges did clear 120 acres—all but 30 acres of their main farm. Now, he looks back with regret: "It lowered the water table and ruined the air. Trees supply your air. It was a big mistake." That night he proudly showed off the stands of trees replanted since then and the way he had cropped so as to prevent erosion. He is, in his own words, "a conservationist."

They chose farming because it is the work and life they desired, and it has not let them down. Over the years it has been a good way to make a living. They have made enough to live comfortably and to have the little things that are important. They were able to help their son go to college and get his start in farming. They have had a chance to work side by side, building their own future. Each was quick to point out how much the other helped in little ways to make the farm a better place. "Both of us like things looking good. We like things to be mowed and kept up, painted, and clean and neat. I couldn't have chosen a wife that was any better. She's the world's best." During the years when they were both in public work, life was too filled with too many things that had to be done. They missed the time together that they cherish now. They tell us more about their life together as they sit side by side on the couch, holding hands.

They see farm life changing, and not necessarily for the better. He recalls how it was when he was growing up:"It was all fun then. You didn't have to think about making a living like they do now. It was fun and I enjoyed it. We really enjoyed life. It wasn't no hassle like it is now. You didn't have to go at it as hard. There was more relaxation to it." Despite the changes, the worries, and the greater effort required to simply get by, they cannot imagine not living and working on a farm. And they see their son continuing the tradition. He will take over the family farms and, like his parents, make a life in farming.

THE END OF A TRADITION

The Lasts operate a diversified 1,900-acre farm raising tobacco, hay, cash crops (corn, wheat, and soybeans), hogs, and beef. Raised on a 50-acre farm, he came from a family of seven brothers and one sister, and he is the only one to make a living in farming. She was raised on a 150-acre

farm, and her parents raised cattle, corn, hay, and for a few years, tomatoes. Both of their fathers also had full-time public jobs.

After a tour of duty in the service, he came home, married, and got a
public job. They lived in town. But he had bought a "a red-belly Ford
tractor to play with," and he would plow up the gardens of his town neighbors. One day a retired farmer living in town told him he needed a little
farm and offered to carry his paper for five years: 50 acres for $6,000,
$1,200 down and $1,200 a year. For seventeen years they both worked
full-time public jobs and farmed their growing acreage. By the time they
quit their public jobs they had about thirty sows, thirty head of cattle, 4
acres of tobacco, and more than 350 acres of row crops. They had to
make a decision. Juggling full-time off-farm jobs and a large farm operation demanded too much time, "We just never did have a minute." So
they made their final choice, to farm full-time.

They never really intended to farm on the scale on which they currently operate. Being from farm families, they both wanted "just to have
a little farm, a place to piddle around." Yet now they operate a farm that
is nearly eight times the size of the average Kentucky farm. They do it
with the help of one full-time worker, who has been with them for seven
years, and two part-time hired hands. They see themselves in partnership
with each other; the farm needs both of them to survive. At first it needed
both of their off-farm incomes, but now it needs their joint labor.

Like many farm women when asked what they do on the farm, Mrs.
Last says, "I don't do a lot." But then she starts to list what she does. She
farrows the hogs; sets, harvests, hangs, and strips tobacco; helps plow
fields; does a bit of the bushhogging; takes crops to the market; helps
move equipment (which is a big job when you are operating five farms
several miles apart); and gofers. Mr. Last listens a while and then jumps
in to say, "She's just as valuable to me as any man would be."

They had hogs until the last few years, until the subdivisions pushed
out toward them. You have to see the green in the smell of hogs to appreciate them, he says; but their neighbors do not. Also, as the size of the
crop operation expanded and new farm parcels were added, the time simply was not there to "fool with the hogs." And so the hogs went, but both
talk about their desire to build a finishing floor and get back into raising
market hogs.

The Lasts have been farming for more than twenty-five years. During
their lives they have seen the total transformation of family farming—from
farms small enough that family labor could work them to multithousand-
acre operations; from hard physical labor to machines that can do in an
hour what used to take a family a day or two; from a pace of life that was

in tune with the cycles of nature to a helter-skelter life synchronized by the time clock and the due dates on notes; from a time of "pay as you go" to a time of debt; from a time when tomorrow's choices and the future were certain to a time of risk, insecurity, and uncertainty. They have grown up and lived their married lives through these changes. They have seen the changes, experienced them, and understand their meanings.

"There is a lot of mental anguish with farming. I'm fortunate in the fact that I've gone through two generations, the old and the new. I look back and think about all that they had to do. They didn't owe anything, but that was because there wasn't anybody that would loan them anything. You might have a little grocery bill run up at one of the country groceries. You didn't owe the banks a bunch of money because they wouldn't loan it to you to start with. The only thing that Daddy had to do was raise the tobacco to have money for Christmas and to get us through the winter. We milked some old cows and got a check every couple of weeks. That kept us in lunch money or what clothes that we bought. I had mostly hand-me-down clothes. He basically raised enough crops to feed the cows that we milked and feed the horses. We had big gardens. Most of our food Momma canned. I don't know how old I was before I ate a hamburger. We virtually didn't have beef. We had bacon and chicken and a little fish if you caught it out of the pond or creek. The only sweets you ever had was when Momma would make a cake. She'd make a cake for every one of our birthdays. That's all we had growing up, but that's not saying that it wasn't a lot.

"But they didn't have the mental pressures on them back then that we do now. It's not just a matter of providing, but it is a matter of paying big sums of money to banks and stuff like that. This is what really bothers you. You say, 'Gosh, how am I going to do this, this year?' Most of the times it always comes some way or another. We talk to each other. You just sit and talk a while and cry on each other's shoulders and pat each other on the back. We're here for one another."

They look around and they see few young couples choosing farming as a business. The amount of capital necessary to start is so large that few can begin even on a small scale today unless they have family backing them. They are beginning to think about retiring. But who will take over their farm? They have a daughter whose husband helps "some" with the tobacco. "But as far as is he going to take over, no. There'd be no way. He's interested, but not so much that he's interested in working like we have to work." And so, "There's nobody who wants to step in and really do what my husband does, keep up with everything that he keeps up with." They look around them at the prosperous farm that a quarter-century of hard work and tears and laughter and companionship have built. "I don't know who's going to be farming after us. It's sad and it kind of scares you."

And so, the circle closes.

REFLECTIONS ON FAMILY FARMING AND FARMING FAMILIES

Farm life is about making trade-offs, constantly adapting to a changing world, and learning to tolerate the unexpected and the inconsistencies in life. It is a crucible of challenges that shapes and strengthens families, or breaks them with stress. In the stories told by these farm families it is clear that farming is both a business and a way of life. It must be, because those who pursue it only for the potential profits or economic gains will survive neither the endless cycles of booms and busts nor the ever narrowing financial returns to their investments.

Farming families are willing to tolerate very narrow profit margins and frequent losses for many reasons. Some, especially the older couples, have enough money to be comfortable, and they have a sense of accomplishment from doing well at what they know how to do. Others are making enough to get by, and although they would like to be more economically secure, they do not think they can get an adequate return for all their investments if they sell out; and even if they did, what else would they do? Still others hold to the wellspring of their faith in the land and their eternal optimism that times will, as always, get better. After all, times have been worse and they have survived. Finally, a thread that runs through all these stories, sometimes clearly stated and more often just hinted, is the understanding that you persist because the farm and the land and the family are one.

In a sense the commitment to the farming way of life is what has tied families to farming despite the economic and emotional roller-coaster ride it offers. The problem today, as so many of these couples note, is that the conditions of the business of farming—the need for off-farm jobs, the larger scale of operations, the heavier debt burdens, and the greater dependence on international market conditions—have begun to erode those noneconomic aspects of the life that once tipped the scales toward commitment. There is less time to spend with family; you cannot work beside a spouse who spends the day in a public job and then rushes to the fields until after dark; you have greater standard-of-living expectations for yourself and your children; and your independence is continually worn down by government regulations and programs. As so many couples said, the fun and the small pleasures have gone out of farming.

What happened over the twentieth century to family farming? Many feel that farming as a business no longer rewards the behaviors—hard work, land stewardship, self-sacrifice—that once were the basis of success. Many also feel that the nation no longer respects the values that have been the foundation of farming—independence, self-reliance, community. We all make compromises in our work and personal lives. Simply consider the independent insurance agent who finally goes to work for a major company because the costs of doing business on his or her own are too great.

The growing unease among farming families is that they are being forced to make so many compromises that they are losing control over their lives. Each compromise forces a reshaping of their lives. Each compromise forces a redefinition of their expectations and their dreams. Government agricultural programs illustrate this process well. To participate in these programs, you must make some compromises. You trade off a measure of control over your own farm operation in exchange for the "security" that comes with the program benefits. But how can you weigh the merits of this trade-off when you can never be sure that the benefits offered in February will be those delivered in August? The fluctuations in government policies undermine any confidence farm families may have in their own decision making.

There is a lurking sense of betrayal. Government programs may not function as promised. The land may not produce as expected, despite giving it all that you have. The market price may plunge the day before you sell. Events and situations thousands of miles away, in places you cannot spell, seem to have more influence on your success or failure than all the planning and work you commit to your farm. Or a child may not step forward to assume the operation as you had hoped and dreamed. You find yourself involved in "what if" and "if only" conversations with yourself and others, sometimes for real, and sometimes only in your mind as you sit on your tractor, take a crop to market, or sit with a sick animal in the quiet of the night. And you mourn your lost dreams.

Yet, despite the uneasiness and the disquiet, farm families and family farming persist. There is strength drawn from working the land together, as husband and wife and as family, and reaping the best harvest ever. There is the sense of wonder, as Barb notes, when a child or a member of the extended family says, "Yes, this is the life and work that I want." There is the satisfaction and self-esteem from seeing what you have created with years of hard work and remembering where you began. There is the confidence that comes from surviving the worst that nature or the marketplace can throw at you, and the hope that things will be better.

Our tendency to view farming and farm life through the images of popular myths has done poor service both to society and to farm families themselves. Our "halo myths" color farming in romanticized images that highlight only the positive aspects, and our "cloudy myths" cloak farming in images that highlight only the dark side of farm life. Neither alone leads to a true understanding of contemporary farm life, just as an emphasis on only the positive economic advantages of being a physician blinds us to the long hours and the stress of life and death decision making. Family farming is a bit of all of these myths, but it is also more. Farming families are families with all the idiosyncrasies and passions and loves and hates and joys and tragedies of other families. Family farming is a business with all the risks and hard work and compromises and stresses and government regulations of many other businesses. They are the same. They are different. That is why these stories echo in our hearts and souls.

Epilogue

EQUINOX

Come in October, friend,
to Basin Spring. Rise early
or stir long and late
in Grandmother's high oak bed,

read and dream all day
or sit in the swing and listen
to summer's last crickets.
The cats will keep you company
as leaves trickle down

until I am home. Then
we'll take to the fields
gather pears and bouquets
   of purple ironweed and goldenrod
   for antique Mason jars,

hike to the spring, and drink
long and cool from the ancient underground,
and talk of you and me and all our years
breathing October
   under lavendar skies.

And when the screech owls begin
their evening dirge, we'll turn home
build a fire in the drumstove
and warm ourselves
with hickory and cider,

dine on biscuits with old ham
   to the strings of Bill Munroe
linger by the stove until stars crackle
    through antique window panes
    lighting the way upstairs.

Dream under your blankets
of births and deaths
   in these tall rooms
of weddings and funerals and
    little feet that have climbed the stairs,

a hundred and ten years
of Christmas mornings
Easter Sunday egg hunts
birthday parties, picnics in the grass,
raking leaves and snowed-in winters.

The leaves will lull you to oblivion,
cattle lowing to their calves
night lives, old loves
October—stirring the life
of Basin Spring.

BASIN SPRING FARM

. . . it begins in childhood, in the heart of a child, with a dream, with the love of a dream, a dream that lingers in the soul and mind and fingertips until all you can do is lay claim to it.

It is a dream of seasons, of stacking wood in January and feeding calves in the frosty February dawn, of sowing greens in early March and hearing April's bullfrog chorus at the spring. It is fencerows thick with May honeysuckle, the fields and trees alight with June fireflies. It is cutting hay in July and cutting tobacco in August, sowing the tobacco patch on a warm September morning, gathering pumpkins and flower seeds in October, stripping tobacco in a November drizzle, cutting a Christmas tree that sweeps an eleven-foot ceiling in a farm house where Christmas has lived for over a century.

It is a dream of history, a dream of love, a dream that stirs when a calf stands to nurse, when spring peepers begin their annual chorus at the spring, when you survey a perfect stand of corn, when fireflies light the fields and trees, when smoke curls into the air with the first fire of the season. Because we are what we love. And because somewhere . . .

Granddaddy is baiting a fishhook by the pond for his boy and girl, and Grandmama is icing fresh banana cake with brown sugar frosting. Chickens are clucking in the henhouse and dogs are napping in the sun. A boy and his dog camp on a shady creek bank, and a little girl plays with kittens in the loft.

So begins the legacy of a farm.

Index